Embodied Sporting Practices

Also by Kath Woodward

WHY FEMINISM MATTERS: Feminism Lost and Found (*with Sophie Woodward*)

SOCIAL SCIENCES: The Big Issues (*2nd edition*)

BOXING, IDENTITY AND MASCULINITY: The 'I' of the Tiger

QUESTIONS OF IDENTITY (*edited*)

THE NATURAL AND THE SOCIAL: Uncertainty, Risk and Change (*co-edited*)

IDENTITY AND DIFFERENCE (*edited*)

THE GENDERED CYBORG (*co-edited*)

UNDERSTANDING IDENTITY

Embodied Sporting Practices

Regulating and Regulatory Bodies

Kath Woodward

Open University, UK

First published 2009 by
PALGRAVE MACMILLAN

Palgrave Macmillan in the UK is an imprint of Macmillan Publishers Limited,
registered in England, company number 785998, of Houndmills, Basingstoke,
Hampshire RG21 6XS.

Palgrave Macmillan in the US is a division of St Martin's Press LLC,
175 Fifth Avenue, New York, NY 10010.

Palgrave Macmillan is the global academic imprint of the above companies
and has companies and representatives throughout the world.

Palgrave® and Macmillan® are registered trademarks in the United States,
the United Kingdom, Europe and other countries.

ISBN-13: 978–0–230–21805–5 hardback

This book is printed on paper suitable for recycling and made from fully
managed and sustained forest sources. Logging, pulping and manufacturing
processes are expected to conform to the environmental regulations of the
country of origin.

A catalogue record for this book is available from the British Library.

A catalog record for this book is available from the Library of Congress.

10   9   8   7   6   5   4   3   2   1
18 17 16 15 14 13 12 11 10 09

Printed and bound in Great Britain by
CPI Antony Rowe, Chippenham and Eastbourne

For Steve, Richard, Tamsin, Jack and Sophie and my sister Sarah

Contents

1
Introduction: Regulating Bodies, Regulatory Bodies

This is a book about embodied sporting practices. The centrality of the body in sport is often taken for granted, but in this book I seek to unravel some of the diverse and complex meanings and practices of bodies in sport. Although bodies are pivotal to what constitutes sport and to how it is understood and experienced, it has not often been sporting bodies that have been cited as the main source of empirical or illustrative material in the development of theories of the body and embodiment. Bodies are material; they include the physical bodies that are involved in sporting practices and the organizational bodies which regulate what is called sport. These bodies are also physical in the sense that they are material and involve social institutions, but the living human body has been seen as distinctive because, unlike, inanimate matter, the living body has some notion of consciousness or intentionality attributed to it. I focus upon the notion of embodiment, rather than 'the body' or even 'bodies' because embodiment always involves a self who is embodied and a self which cannot be disentangled from its corporeality. Embodiment challenges the division between subject and object. The idea of embodiment draws upon phenomenological accounts in which action and behaviour cannot be conceived of as outcomes of will which is then executed by the body. The concept is not without its problems, as I shall demonstrate, but it remains useful because of its incorporation of self and corporeality and of being, doing and thinking. Bodies and selves are inextricably linked, with the body that each of us inhabits offering both possibilities and opportunities and limitations and constraints. To begin with 'the body', or even 'bodies' in the plural, which might encompass more of the diversities and multiplicities of those who are embodied, suggests being outside looking on, as if the body is what is left after the self has gone. In everyday usage, 'the body' might even refer to the dead body. I do, of course, use 'the body', which appears frequently in the literature, but I want to state at the outset that my concern is with embodiment, and my argument is underpinned by the inextricable elision of body and self and the materiality of embodiment. The embodied selves in this book are made and remade through the sporting practices with which they engage and through the diverse regulatory apparatuses of sport.

This book is inspired by reflections upon the specificities and materialities of sport, which have arisen from the work undertaken, firstly, for the *Economic and Social Science Research Council* and subsequently for the *Arts and Humanities Research Council* on questions of equality, inequality and diversity in sport, through its embodied practices, organization and representation. This has been through broadcasting, in my most recent work using the example of the BBC World Service. This research, which has combined discursive analysis of the media in reproducing meanings about sport for the *Tuning In: Sport Across Diasporas at the BBC World Service* and interviews with activists in the field of diversity politics, for *Re-configuring Footballing Identities*, highlights the intersection of representations and embodied selves, which underpins the theoretical concerns of this book. It is this intersection which also presents one of the intellectual dilemmas of exploring bodies in sport, namely the relationship between representations and corporeal, lived experience.

Sport presents its own problems, with a tradition of separation, especially of mind and body. Bodies and selves are, however, inextricably intertwined, although sport can demonstrate spectacular feats of what in common parlance is called exercising 'mind over matter'. My purpose is to focus upon the embodied and enfleshed subject and to explore the centrality of embodied selves in sport, using the vehicle of sport and sporting practices to consider some of the ways in which bodies and embodiment can be theorized in contemporary culture, without resorting to binary logic and by accommodating the intersections between material corporeality and the symbolic and social systems through which embodied selves are made and experienced. The bodies in this book are selves who make and remake experience and selfhood through a range of practices in the field of sport, which is shaped by these embodied selves who participate in such practices and by the regulatory bodies which constitute the organizational apparatuses of sport. Enfleshed bodies are endowed with more status than Judith Butler's (1993) anatomical bodies. These are R.W. Connell's (1995) living breathing bodies which are born, reproduce, lactate, die and, in sport, engage in a range of body practices involving bone, skin, muscle and brain with a nervous system that experiences pain, excitement, distress and a whole range of feelings, emotions and affects.

Sporting body practices and organizing bodies

Bodies matter in sport and sporting bodies are material. The materiality of bodies involves a complex interplay of socially constructed, symbolic and affective forces and is bio-social. Bodies are biological and anatomical and social and symbolic and include contradictions, however emphatic the distinctions that are made in the field of sport. Sports are organized and classified according to what people do with their bodies by moving through space in a variety of ways – running, jumping, throwing, punching, dodging and the skilful manipulation of sporting artefacts and equipment. The kind of body one has and its size are determinants of which competitions one can enter and even, historically, which sport can be played. Sports are very commonly organized into women's sports and men's sports; gender, based on outwardly observable physical and more recently genetic differences, has provided a key criterion of classification for inclusion and exclusion and women have been prevented from playing some sports because they are women. However ambiguous and fluid gender categories may be in some areas of contemporary life, in sport there is a strong adherence to binary differentiation into categories of women and men. Racialized characteristics, similarly, have provided criteria of exclusion from sporting competitions and participation, with bodies being key markers of difference.

Sport measures and quantifies bodies, sometimes formally as part of its rule governing system, as in boxing, into weight categories based on measurable body mass. Sporting practices always involve physical activity of some sort and it is very often the limitations of the body and the damage that is incurred that prevents or constrains sporting success. Athletes often push themselves physically to the limits of physical endurance and competence, even though the damage so incurred can be disastrous. A career can be ended by injury to a single limb. Thus the field of sport offers a particularly productive site for the exploration of embodiment, body practices and embodied selves. It may be assumed that bodies are central to sport but how are these embodied selves regulated, classified and organized? How do they regulate themselves? Bodies in sport and sporting bodies cannot be understood simply, or as separate categories, as if bodies were not inextricably connected to the selves who inhabit them and the social worlds in which sport is experienced. Sport offers a significant site of the intersection between the lived experience of actual bodies and discursive systems of representation. The centrality of bodies and body practices extends beyond the corporeal and includes the bodies that govern sport and establish its rules and regulations, both formally and informally. Sport is part of a wider social world. Sporting bodies are regulated by those who engage in sporting practices and by the diverse regulatory bodies of the social world. These regulatory bodies range from governments and the international and national sports' bodies through voluntary associations and the apparatuses through which sporting practices are routinely enacted and experienced at the local level. Which embodied selves are permitted to participate and which are excluded have been determined by the formal and informal rulings of the various bodies which regulate sport. Different axes of power intersect through the controlling elements of modern bureaucracies of which sport has many with more fluid principles found in organizations that challenge orthodoxies, such as cultural movements that struggle against social exclusion manifest in racism, homophobia or based on gender, disability or ethnicity, as more loosely formed group resistance.

Governments have long targeted the body as a means of creating 'good citizens' as expressed in the 'healthy mind healthy body' dictum. Sport is seen as a healthy physical activity which promotes a fit active body and thus a fully integrated and participating citizen. Boundaries are blurred between the fit, disciplined body and the active citizen. Bodies are also the target of intervention, for example, through the multiple bodies of governmentality, because of the assumption that 'we are our bodies' (Bourdieu, 1992) and that citizen selves can be transformed and recreated through body practices.

Sports heroes and celebrities occupy a cultural terrain in which the media play an increasingly powerful part creating role models and embodied identificatory figures. In this book regulating bodies extend beyond the governance of sport and encompass its governmentality. The media are included in the apparatuses through which embodied selves are made and remade. The sport, media, sponsorship nexus presents a mechanism that dominates the representation of bodies and embodied selves. This book looks at the relationship and connections between the two; between the regulatory bodies of sport and the embodied selves who engage in sport through sets of embodied practices.

The embodied practices of sport are also sources of enormous pleasure for those who participate and for those who are spectators. Sport is fun. It is also challenging and frequently involves competition, but to engage in body practices that extend the possibilities of corporeality and which permit the deepest enjoyment of physicality, as well as the discipline required, raises sporting embodied practices beyond the banal and the routine. Sensations, sensibilities and sensuality are all implicated in sporting embodied practices. Bodies are transported by different pleasures, whether watching or participating. There is no small element of sheer joy in sport that is sometimes obscured by routines and even more by regulatory practices.

Thinking the body

Bodies are not only central to sports performance and achievement, thinking the body in sport is also deeply problematic and indicative of the wider intellectual terrain of theorizing bodies and embodiment. This book is about some of these problems and aims to make a contribution to the theoretical literature on bodies and embodiment using the empirical vehicle of sport. Critical theory emerging from other empirical fields has been applied to sport, but there has been less two-way traffic with theoretical approaches arising from the study of sport being applied more widely, as Jennifer Hargreaves (2002) has argued. Sport and sporting practices provide a lens through which to view the tensions, interrelationships and connections between corporeality and the social world embodied selves inhabit. What sort of problems does thinking about bodies generate?

This book engages with the problem of bodies, notably with ways of theorizing the body and embodiment, by examining the relationship between the bodies which take part in sport and the bodies through which these activities are organized and made meaningful, thus bringing together corporeality and embodied selves and the social world of regulatory bodies, which, like bodies, are diverse and characterized by multiple inequalities. Embodied selves are differently lived and differently situated and, especially for those marked by inequalities, a focus upon bodies and body practices might carry the threat of reductionism and of an overemphasis on physicality. Feminist critics have been troubled by focusing on 'the body' because of its associations with the fixity of corporeally based understandings of gender difference.

My questions in exploring embodied sporting practices include that of how it is possible to acknowledge and address the materiality of embodiment without resorting to reducing the self to the flesh. If you take on bodies, especially if you focus on bodies, does that mean that that is all there is? To target bodies, even in their multiplicities might suggest a reductionism that marginalizes other dimensions of being human. Being reduced to one's anatomical body is particularly perilous for those who have suffered from social exclusion; the bodies on the margins (Woodward, 2009). Alternatively, an excess of social constructionism could lead to an imbalance that diminishes recognition of the materiality of the flesh and the very real experiences of bodies and their practices. Sport demonstrates well the limitations as well as the achievements of corporeality; sport is marked by the successes of physical excellence, the pleasure of physical activity and the fragility of embodiment and the danger of physical damage that can so often restrict sporting body practices and success.

The book explores and develops understandings of embodiment and the relationship between bodies and selves and bodies and communities, using sport as a site in which, not only are material bodies paramount through body practices, but also the bodies which regulate and organize those living, breathing bodies are crucial. Embodied selves are reproduced and represented through the activities and practices of the wider society and culture as well as through the routine practices, for example, of training and of competition in the sport involved. The book considers the extent to which transformations can be made and bodies and selves reformed through the interventions of the diverse bodies that regulate sport by focusing on different materialities and especially the interrelationship of different kinds of bodies, the corporeal and the regulatory. Its theoretical framework deploys post-Foucauldian approaches to the mechanisms of regulation (Brown, 2003; Burchell et al., 1991a; Markula, 1995; Rose, 1996b, 1999), and examines the limitations of an overemphasis on the discursive, including the rejection of psychoanalytical dimensions of identification associated with such approaches and debates that are central to the development of Foucauldian analyses, including Judith Butler's.

Much of the literature on embodiment in sport draws upon Merleau Ponty's phenomenology. Lived bodies are central to 'being in the world' in the phenomenological sense. This book engages more directly with the materiality of bodies and feminist phenomenological critiques of embodiment (de Beauvoir, 1989; Young, 2005b) and in particular the possibilities afforded by de Beauvoir's reading of the body as a situation as well as being situated which are conceptualizations explored by Toril Moi (2000). Developments of de Beauvoir's work on embodiment offer particularly useful ways of combining the material bodies of lived experience with the social, cultural and political context.

Deleuzian understandings of becoming and the notion of assemblage offer a radical critique which has also been developed in feminist critiques (Buchanan and Coleman, 2000) and particularly those such as Rosi Braidotti (1994, 2002), who combines aspects of the work of Giles Deleuze and Felix Guttari with Luce Irigaray's difference-based feminist critique. Few of these thinkers write about sport but the theoretical frameworks they suggest provide interesting opportunities for rethinking embodiment through the prism of sport. The empirical field of sport is deployed in order to address contemporary debates about embodiment and the status of the body in relation to identification and the ways in which identities are adopted by exploring the productive interconnections between the social constructionism and the materiality of corporeality.

Sports methodologies: Sporting practices

Sport presents a particular field of investigation in terms of methodological concerns, methods and what constitutes empirical evidence and which sources are reliable. Sports studies and the sociology of sport intersect with other forms of sports writing and representation, including media coverage of sports events, literary, cinematic and artistic representations and journalism. The boundary between the best journalism and what might otherwise be classified as literature is not always clear, nor should it necessarily be. Journalistic accounts may routinely be characterized by hyperbole and partisan bias, but sport also attracts some of the best and most well-known writing, with authors such as Ernest Hemingway, George Plimpton and Ring Lardner. Particular sports have literary histories, some more so than others; cricket (Cardus, 1947, 1970; Cardus and Arlott, 1986; James, 1963), or boxing (Egan, 1812; Mailer, 1975; Oates, 1987). Such accounts justifiably are included in the source material, which I draw upon in this book, although not without critical evaluation.

Methodological concerns are of prime importance in relation to the theoretical underpinnings of the research. Theory and methodology overlap. Phenomenological approaches are included in the theoretical discussion, especially of different understandings of embodiment and in the conceptualizations of lived bodies and bodies as situations. Phenomenology, with its powerful stress on the testimonies of the subjects of research and on their perceptions of their lived experience has long informed ethnographic studies, for example in anthropology where it has even been claimed that phenomenology constitutes an albatross (Ilde, 2009). However, I suggest that moving into post-phenomenology is not the only solution. The perception of research subjects is not a transparent reflection, nor is ethnography the only possibility. Meanings and practices in sport are made and remade through a number of different genres and fields, including material bodies. It is not necessary to reject the potential of phenomenology, but rather it is necessary to be attentive to the relationship between the researcher and the subjects of research and the diverse materialities that constitute embodied experience.

Reflection upon methodologies and methods is underpinned by consideration of the role of the researcher and the processes through which knowledge is produced and the situatedness of the research process (Haraway, 1991; Stanley, 1997) and as applied to the body practices of sport (Wheaton, 2002, 2004; Woodward, 2008a). Participant observation and ethnography have frequently been used in work on sport and such approaches are particularly attractive because of the embodied practices with which athletes, at whatever level engage. The immersion of the researcher into the body practices of the chosen sport presents great advantages including insights into the lived experience of the sport. Such methods have often been used, in boxing for example (Wacquant, 1995a, 1995b, 2004), as a means of accessing insider knowledge through being in the material, enfleshed world of the boxer. A critical approach to methodologies and to feminist and feminist-inspired critiques to inform the analysis, especially in terms of situating the researcher and acknowledging the situatedness of researchers and the subjects of research (Woodward, 2007a, 2009).

Those who conduct the investigations and generate knowledge are also complicit in the processes through which knowledge is reproduced. Increasingly, researchers whose work on sport draws upon feminist perspectives, including those, such as Free and Hughson (2003) and McKay, Messner and Sabo, described as 'pro-feminist' (2000). Such work engages with masculinities in a way that makes men visible as men, rather than assuming men as the homogenous, non-gendered norm of humanity, moving beyond the female/male binary and putting women into the discursive regime of sport. I would include the lived body of the researcher too.

Not only work carried out by women, but research within an avowedly feminist framework, has been subjected to criticisms of excessive ontological complicity and on occasion of essentialism, for example in the criticisms levelled at French feminist work on embodied gender difference such as the earlier work of Luce Irigaray, who has argued that the notion of gender cannot embrace the specificity of women's embodiment (1991). She demands the recognition of embodied gender difference and presents a powerful critique of the lack of recognition of women's bodies and women's lives in Western culture. This is one aspect of the criticism that has been made of feminist work in relation to its epistemology. The other stresses methodology, for example where the starting point of research is women's experience (Smith, 1997; Stanley and Wise, 1993). In particular, feminist standpoint epistemology, a hotly contested conceptualization, has been criticized, for example, by Wacquant, as holding 'that women's subjugation puts them in a privileged position to produce true knowledge' (1993: 497). Similarly, the development of Chicago School ethnography, involving the participation of men in sports research, might suggest a similar, albeit largely unacknowl-edged, privileging of the researcher's position (Hammersley and Atkinson, 1995, 2005).

Sport is physical for practitioners and for spectators; they are drawn into the spectacle and, as Vivien Sobchack argues in relation to film (2004), spectatorship is a corporeal practice in which all the senses are implicated. The researcher too is physically situated, although, if not a participant, at more of a distance. Although in less dramatic an experience, the researcher, for example, in the gym or at the game, witnesses the regimen of physical exertion, which nonetheless can be powerful. You can smell, hear and see the experience of embodied sporting practices. Spectators' reactions are also physical sensations. In the case of boxing, we are reminded of the flesh at every turn. Joyce Carol Oates suggests this is embedded in the genealogy of boxing (Oates, 1987: 42).

What underpins my interest is the tension implicated in the sit-uation of the researcher. The insider/outsider dichotomy and that between objectivity and subjectivity on which some of these dilem-mas are predicated are based on far too crude a polarization. The research process can never be totally 'inside' or completely 'outside', but involves an interrogation of situatedness and how 'being inside' relates to lived bodies and their practices and experiences. There are myriad ways of being 'inside' sport, although actually engaging in the sport physically is the most dramatic. As the cricketer and sports writer Ed Smith, whose work crosses some of the journalistic, prac-titioner and literary boundaries, notes of the relationship between sport and life; there are multiple ways of 'doing' sport and of it being part of your life (2008). However, it is possible that the more one is immersed in it, the less one recognizes the specificities of engagement. What is often missing in the debate on methodologies, however, is a recognition of the gendered experience of all the lived bodies involved, however they are classified or classify themselves.

One sport, perhaps because it is so marked by traditional gen-der binaries, which illustrates particularly well the assumptions of immersion and the lack of critical awareness of situation, in spite of its explicit dualisms, is boxing. The majority of ethnographies conducted in boxing have been carried out by men (Beattie, 1996; de Garis, 2000; Sugden, 1996; Wacquant, 2004, 1995a, 1995c) as have the great boxing books (Hazlitt, 1982; La Motta, 1970; Mailer, 1975;

McIlvanney, 1996; McRae, 1996; O'Connor, 2002) especially the biographies of Muhammad Ali (Ali and Durham, 1975; Hauser, 1991; Kram, 2002; Marqusee, 2000; Remnick, 1998). Similarly, histories of the sport and expositions of its social significance (e.g., Sammons, 1988; Early, 1994; Gorn, 1986) have male authors. Joyce Carol Oates's is something of a lone voice in the canon of great boxing books (1987). There has been work done on gyms where women box (e.g., Mennesson, 2000) but women researchers are more likely to conduct research by observation and interview, rather than emphasizing participation (e.g., the research of Lafferty discussed in Lafferty and McKay, 2005).

However, men's boxing along with many sports still invites participation in inclusion in a network of colluding masculinities. To be able to say that you 'hang out' with boxers, as 'one of the guys' supports an identification with a powerful, long-standing version of masculinity. The networks through which the culture of boxing is forged remain reminiscent of the *Fancy*, the term used to describe the amusements of sporting men from diverse social backgrounds, in the Regency period in the early nineteenth century in Britain, especially in London (Egan, 1812). Pierce Egan's pugilistic writing in the *Boxiana* series (1812–1829) described not only the theatricality of pugilism and the spectacle of fights, but also the social networks where identifications with masculinity through sporting, drinking and gambling pursuits at times transcended differences of social class. Boxing is illustrative of the gendered specificities of traditional sporting embodied practices and raises some significant issues about the processes through which knowledge is produced, especially about the relationship between the researcher and the subject of the research. Boxing, perhaps more than most sports, invites such dualisms and privileges notions of inclusion and exclusion as one of the defence mechanisms in play in a sport that is so often under attack in the public realm. In this book I am attentive to such collusions and seek to make them visible.

Few ethnographies of sport by male researchers acknowledge or make visible the researcher's gendered identity and maleness often passes unquestioned (Wheaton, 2002). This complicity does not only apply to ethnographic accounts. It extends into other areas of sports research where there is participation in and collusion with the gendered sporting culture even if there is not physical involvement by the researcher in the actual sporting practices (Messner, 2002). Ethnography is one particularly pertinent approach to research, which may be subject to the question of ontological complicity (Merleau-Ponty, 1962) as is the production of authentic 'truth' or an inside world, sometimes couched in terms of the objective–subjective dualism, which have much wider application outside sports research. In order to explore embodied sporting practices and in particular the relationship between the material bodies that participate in sport at all levels and the regulatory bodies that reconstitute them, this book includes consideration of interconnections between different texts and the relationship between all sorts of bodies.

Chapter summaries

The theme of the relationship between regulatory bodies and the bodies they regulate is explored in relation to different sporting practices through the book. Chapter 2, 'Body Matters', sets out the debates with which the book engages in relation to the 'problem of bodies', and the relationship between the bodies that participate in sport and those that regulate its practices in the context of contemporary debates about 'the body' and bodies. This chapter maps out the debates and some of the history of thinking about bodies in sport. What problems does 'thinking the body' pose? How can embodiment be explained without either limiting embodied selves to the material body or, conversely, to a surface upon which social inscriptions are made? This chapter presents some of the problems of theorizing embodiment that have particular resonance within sport and engages with some of the tensions that persist, notably between representations and corporeality and the points of disruption as well as connection between subjectivity and objectivity. Phenomenological critiques have been influential within the study of sporting body practices and this chapter presents a critical discussion of these approaches, which also suggests ways of reworking phenomenology. The problem of thinking bodies also raises questions about the status of the embodied self in the process of effecting change. My argument suggests how change can be addressed by questioning the idea of a linear narrative or epochal transformations and considering the multiplicities of transformation and the interaction between material bodies and social and cultural practices and interventions.

Chapters 3 and 4 bring together bodies in sport and sporting regulatory bodies in the field of categorization. Chapter 3 focuses on sport as a field of enquiry and as a significant social, cultural, political and economic terrain, starting with which rules and practices make up what is called sport. This chapter looks at what constitute the embodied practices that have been and are classified as sport, in order to set up the book's concern with how interventions in the regulation of bodies and embodied selves work in sport. Subsequent chapters focus on the different theoretical approaches that have been deployed to understand different aspects of the relationship between sporting bodies and bodies of regulation, for example as illustrated by the phenomenological practices of lived bodies and the practices of Foucauldian regulatory bodies.

Chapter 3, 'Sport: Bodies at Play?', maps out the field of sport and engages critically with debates about definitions of sport and sporting practices, the significance of sport as play and the relationship between play, games and contests, by exploring some of the distinctions between pre-modern and modern sport (Giulianotti, 2005a; Guttmann, 2005; Ingham and Loy, 1973, 1993; Loy, 1968). The transition into the period of modern sport is marked by the changing relationship between the governance and the apparatuses of regulation in sport, evident in a range of sports which are covered in the book, such as athletics, boxing, cricket and football (soccer). A focus upon the ludic dimensions of sport and the universality of its rules can obscure the marginalization of some groups and the inequalities that so mark the field of sport and its histories. Sporting bodies are temporally and spatially situated within different fields, ranging from specific bodies, such as the Olympic Committee, the IOC, Athletics Association (AA), the Football Association (FA), the Test and County Cricket Board (TTCB) and the World Boxing Association (WBA), to legislative and quasi-governmental bodies and the media which plays a key role in the making and remaking of sport. The chapter highlights some of the histories of the making of the healthy and 'normal' sporting body and the evolution of regulation in sport, for example in relation to the articulation of gender, sexuality, able-bodiedness and race and the construction of success and failure in sporting practices.

Chapter 4, 'Equalities and Inequalities: Diversity and Neo-Liberal Dilemmas', focuses upon some of the dilemmas and tensions between the inequalities that persist in sport and the opportunities that participation offers. The chapter starts with a discussion of equality within discourses of liberalism and the problems of equality and difference and the move towards the idea of diversity as a solution to the oppositional language of difference. This chapter also addresses the strategies and practices of Foucault's notion of governmentality, for example as applied by Nikolas Rose (1996a, 1999), that have been adopted to promote greater social inclusion and cohesion through sport and explores the measure of transformations that have and have not been made possible. Using the examples of those classified as outsiders on grounds of gender, ethnicity or 'race', especially, theorizing the body can be seen to present the dilemma of reductionism and the threat of being reduced to the body, whereby success or failure is measured in relation to corporeal characteristics. Sports like soccer, cricket and boxing have a long tradition of social exclusion and inclusion and of particular class, 'race', 'ethnic' and gender affiliations which demonstrate the centrality of material embodied differences which inform debates about equality and inequality and social inclusion.

Chapter 5, 'Embodied Identities: Situated Bodies, Bodies as Situations', highlights the centrality of embodiment in sport through the evolution of regulatory practices, for example in the tradition of Juvenal's *mens sana in corpore sano*, which at different historical moments has been applied to different social, ethnic and cultural groups, where the healthy person is constructed as the physically healthy individual. What identities are configured in sport and what problems emerge from seeing sport as a site where such embodied identities are forged? This also raises the need to address some of the problems of 'thinking the body'. For example, the devaluing of the body in relation to the mind can be lived in the experience of social disadvantage which operates in contradictory ways in sport, which has been addressed in phenomenological accounts which challenge the mind–body binary (Bourdieu, 1992; Crossley, 2001; Merleau-Ponty, 1962) and the centrality of body practices in the making of gendered identities (Connell, 1995, 2002; Young, 2005a). This chapter explores the possibilities of different aspects of embodied selves and looks at how actual bodies are lived, situated and at the concept of bodies as situations (Beauvoir, 1949; Moi, 1999). Individual and collective endeavours in sport focus upon the material body. Sport has been seen as an appropriate route for black young men, as in athletics, boxing and the divisions which operate within cricket, which could reinforce inequalities or provide opportunities for agency and resistance (Carrington, 1998; James, 1963). Cricket offers a good example of historical reworkings of class and gender binaries and the operation of racialization and ethnicization (Carrington, 2001; Tomlinson, 2007) as does boxing, through the subtle injunctions of power as well as its more coercive manifestations (Sugden, 1996; Wacquant, 2001, 2004; Woodward, 2007a). This chapter presents the argument that bodies are both situated and are themselves situations, in order to accommodate all elements of the materiality of embodiment, thus combining analyses of embodiment with those of regulation.

Chapter 6, 'Beyond Text: Spectacles, Sensations and Affects', suggests other possibilities for interrogating what Butler calls the 'discursive limits of sex' (1993) in relation to embodied selves. The chapter explores how sport invites spectatorship and very powerful allegiances, for example through the identifications of fans and supporters and through performance and spectacle. Identifications and affiliations in sport can also be understood within the parameters of belonging and becoming (Braidotti, 2002; Deleuze and Guattari, 1987) and seen not to follow a linear path. Meanings about body practices and embodied selves are recreated through routine practices, for example in the gym and the public arena of representation and the media, and this chapter looks at an example of broadcasting which brings together everyday followers and media heroes, using the example of the BBC World Service coverage of sporting heroes. Celebrities (Whannel, 2002; Cashmore, 2004), whether configured as sporting heroes, personalities or villains, are a key component in the production of meanings about embodied sporting identities. This chapter explores the status of such representations and the identifications that are made by examining the phenomenon of the sporting spectacle and the celebrities who inhabit these spaces as well as those who are invisible and under-represented. Spectacles such as The Olympics offer a site at which bodies are central to the political iconography of sport as manifest in some of the big moments at the Games, including the podium salutes of Tommie Smith and John Carlos at the 1968 Games. Media representations include live events, television and film and there are distinctions that can be made, especially between the live event and its representation on screen (Oates, 1987). This is used in this chapter to explore the assemblage of sensations and affects which constitute the embodied self (Sobchack, 2004) and the impact of the affective turn in sport which goes beyond visibility and the visual to include aspects of the sensate and sentient body such as sensations of hearing, touch and smell and emotions and affects.

Chapter 7, 'Beyond Bodies', examines the extent to which body projects and practices facilitate transformations and the degree to which, in sport, such interventions may or may not transgress and contravene regulatory practices. The promise of monsters (Braidotti, 1996) and of cyborgs (Haraway, 1990), which inform sporting practices that range from technologies of health and nutrition, the technoscience of sports science to pharmaceutical interventions, for example in athletics, although there is extensive evidence across a range of sports. The chapter cites the example of the Paralympic sprinter Oscar Pistorius and his status in sport as 'Blade Runner'. Cyborg thinking offers one explanatory framework which has to be negotiated within a framework of ethics (Walsh and Giulianotti, 2006) and a critique of power. This chapter includes the crossing of other boundaries, such as those between human and machine and human and animal (Haraway, 1991, 2007). Sport invites transformations of the body as well as providing a site where the boundaries of the human body can be interrogated and challenged as well as possibly reinstated. Self-regulation and techniques of the self suggest the transgression of body boundaries but also demonstrate tensions between regulatory bodies and the material bodies they target and present only limited acknowledgement of the inner as well as the outer worlds through which selves are made (Butler, 1990; Grosz, 1994). This chapter explores how sustainable the reiteration of fluidity can be in explaining embodiment and the relationships between different materialities. Boundaries can be set through specific assemblages.

Chapter 8, 'Conclusion', returns to the initial 'problem of bodies' and Butler's question about the relationship between material bodies and regulatory discursive regimes and brings together the arguments of the book which extend beyond the constraints of a polarization between social inscriptions and material corporeality, permitting a conceptualization of bodies and selves as assemblages. Bodies are both situations and situated – assemblages themselves and part of the systems which constitute assemblages in the Deleuzian sense. However, there is a self and even a narrative of that self which retains some continuity of energy which is bounded to some extent by whatever changing body the self lives in, in the world. Sporting bodies and sporting practices offer the possibility of transformation and of retrenchment and also highlight the disjunctions and disruptions as well as the continuities between disciplinary regimes and regulatory practices and the material bodies which they target. The conclusion revisits the concept of change and points to a reformulation of the moments through which change can be seen to be effected, both incrementally and in significant ways which can be understood as multiplicity, where change involves movement rather than a simple forward trajectory.

2
Body Matters

Introduction

Bodies matter because we are our bodies and, at a common sense level, the body would appear to set the limits of the self: 'one body one self' (Fraser and Greco, 2005: 12). The assumption that the body and the self are one underpins the feminist-inspired handbook of second-wave feminism *Our Bodies, Ourselves* (The Boston Women's Health Collective, 1978) and suggests that gaining control of your body means gaining control of yourself and that the body is an intrinsic part of the self. The everyday elision of body and self also suggests some boundaries, for example between bodies and between the embodied self and the world outside that self. There is a strong sense in which bodies are experienced individually, not least because pain (Scarry, 1985) is located within a specific body and the body is mortal. The body sets the limits of the self as well as restricting, and enabling, what embodied selves can be and do. The boundaries that are so powerfully challenged in rethinking the body constitute a resilient common sense in lived experience and routine practices. 'The body' and especially the idea of an embodied self also invokes ideas about agency, which could mean that people might have some control over their own bodies and shape themselves through their bodies. Agency is made possible through bodily knowledge and corporeal experience. There is also a sense of heroic endeavour in pushing the body to its limits that has a powerful presence in sport, especially in relation to traditional masculinities, even though such practices might be a terrible cost. To deny the frailty of the material body may court injury and failure.

Such ideas underpin commercial enterprises, especially those linked to sport, health and well-being, concerned with various forms of 'body projects in contemporary cultures which powerfully promote individual responsibility' (Shilling, 1993). Bodies are also big business in late modernity and sport is part of the culture of self-improvement that deploys body projects through which an aesthetic and health-related enhancement of the embodied self is made possible. The perfect body, encoded and represented as slim, fit and healthy is one of the aspirations of ever more unhealthy and overweight Western societies. The routine practices of sport might offer a means of attaining the perfection of the sports star and of the beautiful athletic body, although the big business of sport is more likely to be concerned with sport's commercial possibilities which target spectators more than participants. Nonetheless, bodies lived and represented are central to sport and its practices. This chapter maps out some of the ways in which bodies and embodiment have been theorized in the relatively recent explosion of literature in the academy which engages with the subject, starting with the arrival of 'the body' as a subject of intellectual enquiry and research in the social sciences and humanities. The chapter explores some of the problems that emerge from this focus upon bodies in order to apply the intellectual concerns of conceptualizations that may not have been linked to the empirical field of sport but which nonetheless have significant resonance for the focus of this book.

Whilst it is my concern in this chapter to challenge the notion that there is a unified 'body', which can be separated from the mind or self, or which is undifferentiated, I use the singular noun with its definite article as a shorthand way of expressing some of the history of 'the body' and its different meanings, although much of the literature now insists on the plurality of bodies to accommodate diversity. Retaining reference to 'the body' also demonstrates the centrality of corporeality, as the material, living, breathing body of everyday experience, especially in an exploration of differentiation and inequality. This material body is widely implicated in sporting practices and is the matter of regulation and the target of the regulatory bodies of sport.

Theories of the body: Bodies of theory

Over the last 20 years, 'the body' has become a major concern of academic study in the social sciences (e.g., Butler, 1993; Featherstone et al., 1991; Martin, 1989; O'Neill, 1985; Scarry, 1985; Shilling, 1993; Turner, 1984, 2004 and in collections like that of Fraser and Greco, 2005). Corporeality and visible embodied difference has always underpinned social divisions, but it is only relatively recently that 'the body' has become an explicit subject of research in the social sciences and humanities. Feminism has also taken a 'corporeal turn' (Howson, 2005). Bodies on the margins are the site of the experience of social exclusion, where economic, political, social and cultural inequalities are played out. Those on the margins have powerful needs to address the ways in which corporeality contributes to marginalization and this has been particularly important within feminist thinking (Howson, 2005). The actuality of living in a particular kind of body has been assumed rather than stated, especially when that body is classified as a woman's body. The 'normal body' has long been assumed to be that of an able-bodied, white male.

Sociological theories and what has been called the sociology of the body have engaged with some of the problems of developing a sociological understanding of what at first appears to be a natural phenomenon. An interest in the natural has also been the starting point, for example, for feminists who have been motivated to demonstrate that women's bodies are not fixed by nature and that change which challenges oppression predicated on immutable, natural stasis is possible; the deconstruction of the 'natural' has had a social and often political dimension, from the start by seeking to subvert the constraints of a natural, biological defence of an inequitable status quo.

Theoretical perspectives to study of the body have adopted different approaches, which, although there are intersections and overlaps, can be seen as following particular trajectories. They share a concern with the embodied materiality of what can be seen to constitute the self and an awareness of the importance of the material body. As Dutton argues,

> What subject after all could be closer, figuratively or literally, to our human concerns? The body is the focal point of our individual identity, in that we not only *have*, we *are* our bodies, however

distinct the body may be conceptually from the 'self' which experiences and knows it is by its nature an embodied self, a self whose social identity and whose location in time and space are contained and defined by their individual embodiment. (Dutton, 1995: 11)

The embodied self is temporally and spatially defined and contained as Dutton argues. These constraints also offer something of a challenge with which sporting bodies have to engage. Recent theoretical perspectives on bodies and the body have implications for the study of embodied practices in sport in different ways, although some have particular applications and can be combined most effectively, as I shall demonstrate. These perspectives emphasize four broad strands. First and in a view that is strongly expressed in sociology, cultural studies and feminism, it is argued that the body is not a natural phenomenon but is socially and culturally constructed. The social and cultural terrains are what make and remake bodies. Rather than being limited to a set of body parts, bodies are socially and culturally inscribed. Second, the body is construed as a representation of social relations of power in another view which stresses extra corporeal dimensions of embodiment, but uses metaphors of the body to inform analyses of power, as in the case of the body politic. Such body metaphors have long been deployed in political science and in sociological accounts of social relations, for example in Durkheimian, functionalist explanations of the social organism, where the operation of society is compared to that of the body. A third view stresses the phenomenological experience of embodiment in the everyday world. The last approach, which has been significantly influenced by anthropology and which has particular resonance in the field of sport, looks at the body as a collection of practices or techniques.

Feminist theory has been particularly influential in the development of analyses of the social construction of the body, which can be seen as representing a strand in thinking the body. Feminist theories are particularly relevant to my project because of their central role in the development of critiques of bodies and embodiment and because gender is a key aspect of differentiation in sport which is often based on assumptions and categories of physical attributes especially in relation to gender. Feminist theories have the advantage of encompassing political and ideological underpinnings which mean that explanation has often been motivated by active engagement in body politics. Chronologically, second-wave feminist thinking can be seen as deriving from Simone de Beauvoir's *The Second Sex* (1989 [1949]) and her much quoted, apocryphal claim that one is made and not born a woman. Thus what is taken as sexual difference is the result of social, cultural and psychological processes which construct femininity as a set of essential attributes possessed by those who are classified as women. De Beauvoir's work inaugurated a research tradition concentrating on the social production of differences in gender and sexuality, which created a range of feminist theories of the body mostly based on social constructionism; the differences between bodies that are categorized as female or male which are assumed as facts of nature are socially produced. At some points in the history of feminist thinking about difference, sex and gender have been separated as distinct categories in order to distinguish between what is biological and what is social. Second-wave feminism in the 1970s was also concerned with interrogating distinctions between sex and gender and in establishing the notion of difference between biologically determined sex and the social construction of gender roles and sexual identities (Oakley, 1976) as outlined by Diane Richardson in the context of developments in women's and gender studies (Richardson, 2008). The concept of gender as centring on socially constituted difference based on hierarchies and inequalities between women and men has had diverse history ranging from the apparent simplicity of the sex/gender binary which could be based on performance in the sociological sense (Goffman, 1959) to the more complex, postmodernist understandings of sexual difference as performative (Butler, 1990). One of Judith Butler's most radical and important contributions to debates about body matters, as well as one with troubling implications, has been her challenge to the sex/gender binary and her argument that sex too is socially constructed (1993). The experience of corporeality has been central to feminist empirical research which has demonstrated the social and political subordination of women which is expressed in psychological depression and physical illness, anorexia nervosa, obesity, and eating disorders (Bordo, 1993) and body image (Orbach, 1985). However, in this book, I argue that feminisms are both constitutive of, and productive within, all the traditions of thought which make up thinking the body, rather than making up a distinct and discrete strand, although they have had particular impact upon social constructionist approaches. Most importantly, feminists have focused on difference and inequality and have provided intellectual frameworks within which difference, however conceptualized, is central and assumptions about homogeneity are questioned.

Second, also in a strongly social constructionist tradition, the body has been seen as part of political discourse as a representation of power and as a cultural representation of social organization. Corporeal metaphors describe the processes of governance, which through diverse mechanisms reproduce bodies. As in this book, bodies are both physical entities and organizational within the governance of sport, an approach which draws strongly on the work of Michel Foucault (1973a, 1973b, 1974, 1977, 1981, 1988a, 1988b). The body enters into political discourse as a representation of power, whereby power is exercised over the body as in Foucauldian bio-politics. Foucault's work has particular resonance for an investigation of the relationship between enfleshed bodies and the apparatuses of governance. I use enfleshed here to foreground the living breathing body that seems absent from discursive approaches which so highlight inscription and social construction. Even if it is a body that is inscribed it is inscription rather than corporeality that matters in such approaches. Post-Foucauldian work takes diverse paths and includes research which focuses on the specificities of *how* bodies are inscribed. Following Foucault, representations of the body are the result of relations of power, which operate diffusely, for example in the relations between women and men (McNay, 1992). Work in sport is expanding because sport offers a rich site for the exploration of the operation of power at different levels, as well as through the micro-levels and visibility of surveillance (Markula, 2003, 2004; Markula and Pringle, 2006).

Links between embodied subjects and the regulatory bodies of governance also foreground gender and draw upon classificatory metaphors of embodied difference; anatomical maps of the human body vary between societies in terms of the dominant discourse of gender (Laqueur, 1990). Bodies and body practices are differently evaluated and encoded across time and space, even in sport (Spence, 2008). In the anthropological tradition (Douglas, 1966) the divisions of the body are used to make moral distinctions between good and bad and in the construction of the 'stranger' or 'outsider'; embodied, visible difference is the mark of otherness. Bodies themselves can thus be seen as markers of pollution; blood can be polluted (Laws, 1990), left-handedness represents things that are sinister and the skin is both a physical and cultural boundary and a marker of inclusion and exclusion (Elias, 1982).

Another approach which addresses the problem of the relationship between representation and discourse and materiality and offers what is probably the most radical rethinking of bodies, which creatively and profoundly subverts dualisms, is the Deleuzian notion of bodily materialism. I have included Deleuzian approaches here as they can be seen as in dialogue with Foucault's work although there are marked points of difference. Deleuze sought to redress some of the excesses of postmodernist social constructionism by positing an enfleshed, vitalistic but not essentialist vision of the human subject. In the Deleuzian version of becoming material, the post-human body, which occupies the space between humans and machines, is an abstract machine which captures, transforms and produces interconnections. Deleuze and Guttari's concept of the 'body without organs' (1987) constituted by energies and flows has been deployed to demonstrate that cyborg subjectivity belongs not only in the realm of science fiction but the blurring of boundaries between human and machine is enacted in multiple layers of experience, from medicine to consumption and from telecommunications to warfare. Deleuze's ideas have been taken up by some feminist theorists, for example Rosi Braidotti (2002) and Elizabeth Grosz (2000), because of the challenge they offer to binary logic, fixity and hierarchical stability, for example of the nature culture divide. Deleuze and Guttari's materiality offers greater emphasis upon energy, productivity and creativity than the constraints of 'docile bodies' that have been seen as problematic in Foucault's earlier work. The Deleuzian notion of assemblage offers an alternative version of the operation of power and of transformation, which is also explored later in this book.

Fourth, Maurice Merleau-Ponty's concept of the 'lived body' as developed in his *Phenomenology of Perception* (1962) has been developed more recently, in the field of sport (Crossley, 2001; Wacquant, 1995a, 1995b) and in feminist phenomenological approaches (Moi, 2000, not in sport, and Young, 2005b, with application to sport). Young is critical of Merleau-Ponty's phenomenology because of its failure to incorporate power relations, especially those based on gender difference, but she takes the lived body as her starting point and engages with phenomenology in her account of the modalities of the female body. Phenomenology opens up ways of engaging with the thinking body and the embodied self as a body subject which is not divided into either mind and body or body and consciousness. In developing the phenomenology of the everyday world, Merleau-Ponty applied Husserl's phenomenology to intentional consciousness within a corporeal understanding of human consciousness, perception and intentionality. The idea of embodiment draws upon such phenomenological accounts, especially of what is distinctive about the human body which 'holds sway in consciousness' (Husserl, 1970: 107), and where consciousness is always experienced through the body. Merleau-Ponty argued that the body 'is not, as it were, the handmaid of consciousness, transporting the body to that point in space of which we have formed a representation beforehand' (1962: 139). 'Consciousness is in the first place not a matter of "I think that" but of "I can". Consciousness is thus a "being-towards the-thing through the intermediary of the body"' (ibid.: 137).

Merleau-Ponty aimed to avoid the binary logic of the division between the subject and the object as expressed in Descartes's *cogito ergo sum*. Merleau-Ponty's idea of the 'body–subject' that is always situated in a social reality has involved a rejection of behavioural and mechanistic approaches and been used to demonstrate that the body is central to our being in the world. Perception cannot be treated as a disembodied consciousness, mind and body merge. Phenomenology has significant methodological advantages for the exploration of lived experience and the expression and inclusion of marginalized voices.

Lastly, in an approach that has been linked to those of phenomenology, human beings are seen as embodied through acquired corporeal or 'body techniques', a concept developed by Marcel Mauss (1989 [1938]) to describe how people learn to manage their bodies according to social norms. Mauss's anthropological legacy suggests that we think about the body as an ensemble of performances and has been developed by Pierre Bourdieu in terms of two particularly influential concepts. One is hexis, which refers to bodily comportment such as gait, gesture or posture and the other is habitus, which refers to the dispositions through which taste is expressed in the habitual way people act. Bourdieu employment of these concepts in his empirical study of French society in *Distinction: A Social Critique of the Judgment of Taste* (1984) has been enormously influential and has had a significant impact in the field of sport because it focuses on the material body and its practices and its representation as well as including the distinctions of taste as applied to sport within its empirical remit. The body is invested with symbolic capital whereby it is a corporeal expression of the hierarchies of social power. Thus the body is permanently cultivated and represented by the aesthetic preferences of different social classes whereby, in the example of twentieth-century French culture, mountaineering and tennis were associated with the middle and upper classes, whereas the working-class sports of wrestling attracted and produced an entirely different body and habitus and investment of physical capital. Bourdieu's work has been influential in studies of sporting habitus notably in boxing from the Bourdieu scholar and Bourdieu's co-author, Loïc Wacquant. Bourdieu's work has been deployed by feminists, for example Beverly Skeggs (1997), but, although his own theoretical framework accommodates social class well, he was less successful in his treatment of gender difference, his engagement with 'masculine domination' notwithstanding (Bourdieu, 2000).

These theoretical traditions of thinking about the body are underpinned by two alternatives. They are broadly divided by two views. On the one hand, the cultural decoding of the body as a system of meaning suggests that it has a definite structure existing separately from the intentions and conceptions of individuals. On the other hand, as in the phenomenological approach to embodiment, attempts to understand human practices are organized around the life course of lived experience. The work of Bourdieu offers a possible solution to this tension between meaning and experience and between representation and practice. Bourdieu's development of the notions of habitus and practice in *Outline of a Theory of Practice* (1977) provides research strategies for looking simultaneously at how status difference is inscribed on the body and how we experience the world through our bodies, which are ranked in terms of their cultural capital. This reconciliation of these traditions can be assisted by distinguishing between the idea on the one hand of the body as representation and on the other of embodiment as practice and experience.

In much of the work drawing upon Bourdieu, mind and body have been elided, especially in the field of sport (Wacquant, 1995a, 1995b,

1995c) and further attempts have been made to address the problem of the gap between representation and experience deploying Bourdieu's notion of symbolic capital applied to routine embodied practices. As Bourdieu showed in *Distinction* (1984) the body is invested with symbolic capital in which the body is a living expression of the hierarchies of social power within the habitus of social class. Ways in which the body is permanently cultivated and represented by the aesthetic preferences of different social classes have particular resonance for the study of sport. The different sports that are supported by different social classes illustrate this form of distinction so that boxing is part of the habitus of the working class and golf, of the upper classes.

An overemphasis on understanding the body as a system of cultural representation makes it difficult to develop an adequate explanation of bodies as part of lived experience. Bodily performances cannot be grasped simply as static representations (Shusterman, 1992) and an aesthetic understanding of performance cannot neglect the embodied features of artistic activity. The need for an understanding of embodiment and lived experience is crucial in understanding performing arts and for the study of the body in sport and it is not surprising that Bourdieu's approach has been so important in critiques of embodied sporting practices and in the sociology of sport.

Bourdieu offers one possible solution to the division between the meaning and experience of embodiment or the cultural representation of the body. Bourdieu's development of the notions of habitus and practice in *Logic of Practice* (1992) creates research strategies for examining how, for example, status differences are inscribed on the body and how we experience the social world through our bodies which are ranked in terms of their cultural capital. This analytical reconciliation can be supported by clearly distinguishing between the idea of the body as cultural representation and embodiment as practice and experience. Bourdieu's work does offer a route into synthesizing phenomenological theories, body practices and technologies and social constructionist understandings of embodiment.

Phenomenology and social constructionism

Social constructionism, in its myriad forms, and phenomenology, present distinctive, and in many ways contradictory, approaches to the body, which nonetheless have points of reconciliation and rapprochement. These two traditions present different answers to questions about how far the human body is socially constructed. Social constructionism offers the advantage of stressing the inscription of the body as a constructionist system of cultural representations, whereas in the phenomenological tradition, the 'lived body' is studied through its routine practices in the everyday world of social interaction, which clearly has resonance within the field of sport. Phenomenology stresses the ways in which cognition is always an embodied perception of the world and offers a critique of the dualism of the mind and body, in which body is seen to be passive and inert. However, this permits and even prioritizes a degree of individual agency which postmodernist and post-structuralist social constructionist approaches would seem to deny when bodies are understood as cultural representations of social life. The Foucauldian perspective is less concerned with understanding experiences of embodiment; it does not aim to grasp the lived experience of the body from a phenomenology of the body. This apparent division between the body as representation on the one hand and the body as experience on the other has dominated much of the sociological debate about the body, although more recently there have been many attempts to reconcile these differences, for example drawing upon the work of Bourdieu (1977, 1992, 2000) and especially to demonstrate some of the ways in which representations intersect with experience so that the two cannot be disentangled. More recently, but largely not in the field of sport, there has been interest in Deleuzian solutions to the problems of such persistent binaries, especially in Deleuze and Guttari's challenge to the linearity of dualistic thinking (1987).

Foucault's bio-politics and his historical studies of medicine and sexuality (1973a, 1981, 1988a, 1988b) have been among the most influential in providing a framework for the discussion of the relationship between regulatory bodies and regulating bodies through the interaction between the body and systems of knowledge and belief within the social constructionist trajectory. Foucault opened up new ways of thinking about how bodies are imagined, constructed and represented, although with less emphasis on experience which lies at the core of phenomenological approaches.

Bio-politics and regulatory bodies

Human bodies have become central to economic growth in a wide range of bio-tech industries, including sport. Somewhat paradoxically, the pathology of the human body is itself a productive factor in the economy of sport. Physical failure can be prevented by a variety of interventions and the damaged sporting body is the source of investment in technologies. Because of the financial returns on sports investment and the massive earnings of the wealthiest clubs and athletes, sports medicine offers opportunities for investment in progress and new techniques. Sports medicine and sports physiotherapy are in the vanguard of the development of new techniques. In the wider arena, there is enormous interest in the human genome project and its implications for new reproductive technologies, cloning and genetic screening, are important illustrations of public concern about the social consequences of the new genetics. Bodies in sport also lie at the centre of legal concerns about which interventions are permissible and which are not, notably in terms of performance enhancement through pharmaceutical aids and, more recently, debates about disabled athletes, which are explored more fully in Chapter 7. The cultural, analytical and political problems about how to understand and how to manage bodies are central to the study of contemporary embodied sporting practices.

Developments in bio-medicine demonstrate Foucault's distinction between the study of the individual body and the study of populations, whereby Foucault examined how various forms of discipline of the body have regulated individuals in the anatomo-politics of the human body (1981). Anatomo-politics is concerned with the micro-politics of identity and concentrated on the sexuality, reproduction and life histories of individuals, but can equally well be applied to sporting body practices. The clinical examination of individuals is part of the anatomo-politics of society. The bio-politics of populations used demography, epidemiology and public health sciences to examine and manage whole populations, and sport science is playing an increasingly important role in this field. Foucault's understanding of bodies was organized around conceptualizations of discipline and regulatory controls. Developments in technoscience have created enhanced opportunities for governmentality as a strategy of political surveillance and economic production (Foucault, 1977, 1981).

This has significant applications in the field of sport, first, because of its focus upon bodies and, second, because government of the body is a critical issue in the management and regulation of individuals, as well as populations, in contemporary society (Cole, 1998; Markula, 1995).

Sports medicine not only is a vital part of contemporary discourses of medical science, but also carries important wider implications for sociological studies of bodies and embodiment. The fit, sporting body is also a medicalized body which has become part of the core concern of theoretical sociology, especially in debates about the extent to which embodied selves are self-regulating and the transformation of the interrelationship between the social and the natural, often translated into that between corporeal, embodied selves, regulating bodies and social constructionist understandings of the self and the body, shaped by bodies of regulation.

The other bodies with which this book engages are the bodies which regulate sporting embodied practices. Such bodies have a dominant presence in the socio-cultural terrain within social constructionist approaches, especially those which draw upon Foucauldian perspectives. In all social constructionist approaches, institutions and cultural systems play a key role in the recreation of bodies and embodied selves. In Foucauldian theories, these bodies are implicated in the micro-processes of administration and control within which self-discipline and control are located; the bodies which constitute Foucault's idea of governmentality. Governmentality provides an integrating strand of social, cultural and political practices by which the self is constituted through disciplinary regimes. The micro-power relations, through which bodies are regulated by local institutions and authorities, constitute an ensemble:

> The ensemble formed by the institutions, procedures, analyses and reflections, the calculations and tactics, that allow the exercise of this very specific albeit complex form of power, which has as its target population, as its principal form of knowledge political economy, and as essential technical means apparatuses of security. (Foucault, 2001: 219–220)

It is these complex micro-power, regulatory relations which characterize sport in modernity. The state intervenes in the promotion of sporting body practices in which people might well want to engage anyway, but the state and the intervention of myriad other regulatory bodies promote and discipline what counts as sport and how it is understood. Thus, Foucault's notion of bio-politics encompasses the practices of the embodied self and the practices of the regulatory bodies which make up the apparatuses of governmentality.

Body practices: Bodily reflexive practices

While much of the sociological and anthropological tradition explores the body as a symbolic system, embodied selves learn and practise a variety of cultural practices that are necessary for everyday activities through body techniques (Mauss, 1973). For Mauss the body was a technical instrument which is organized in different ways in different cultures (1979). The anthropological assumptions emerging from Mauss's work were developed by Bourdieu through the concepts of hexis and habitus in which dispositions and tastes are organized. Bourdieu's work has led to some innovative approaches for understanding the relationship between identity and embodiment in sport (Wacquant, 1995a, 2004). Physical comportment clearly involves body techniques and reiterated actions through which identities are forged and enacted.

Accounts which are critical of the excesses of social constructionism have seen this focus as offering a route through which corporeality can be reinstated and understood as part of the making and remaking of the self. If social constructionism is too disembodied and writes off the body, how can bodies be recovered without claiming that, somewhat superficially, social meanings are determined by anatomy or prioritizing a biological base? A deconstruction of body practices which is reformulated as body reflexive practices, drawing upon Mauss's concept of body techniques, affords some advance on the binary logic of this dilemma.

Some approaches point to other limitations of social constructionism and provide alternatives that stress the inextricable mix of the body and the social; the body is always present in the social and the social is always present in and mediated by the materiality of the body. As R.W. Connell argues,

> Bodies cannot be understood as a neutral medium of social practice. Their materiality matters. They will do certain things and not

others. Bodies are *substantively* in play in social practices such as sport, labour and sex. (1995: 58)

Connell's example in this discussion in 1995 is of masculinities, which has the advantage of focusing upon gender, which forms a key organizing principle in sport; such 'body reflexive practices' are described as

> not internal to the individual. They involve social relations and symbolism; they may well involve large-scale social institutions. Particular versions of masculinity are constituted in their circuits as meaningful bodies and embodied meanings. Through body-reflexive practices, more than individual lives are formed; a social world is formed. (Connell, 1995: 64)

Through body reflexive practices, bodies are thus drawn into history:

> without ceasing to be bodies. They do not turn out to be symbols, signs or positions in discourse. Their materiality (including material capacities to engender, to give birth, to give milk, to open, to penetrate, to ejaculate) is not erased, it continues to matter. (Ibid.: 64–65)

In this conceptualization, bodies are both agents and objects of practice and it is through the practices of the body that structures are formed. The body's materiality enables and constrains what the social world can impose or influence. Connell argues against the notion that gender differences are determined by the bodies people inhabit and for the notion that, for example, masculinity is mutually constituted by the bodies that act in the world and the social world which would be formed by and make the practices that are construed to be masculine.

Gendered bodies: Feminist contributions

Feminist theories of the body and embodiment have not only been central to the development of conceptualizing and understanding bodies and embodied selves, but also been both creative and innovative in deploying other approaches and bridging seemingly different and incompatible theoretical perspectives. 'The body' and, more especially, bodies have always played a key role in both women's and gender studies. The body and ways of thinking about the body have an unsettled history in feminist thought, however. At times it might have appeared that, by focusing on the gendered body, feminists were falling into the trap of associating women with 'nature' and by implication linking men to 'culture', with all its connotations of superiority and rationality. There is sometimes a slippage between biology and embodiment, with corporeality being linked to biological imperatives. As Steven Rose has argued, the cultural devaluation of the body has arisen historically from its associations with biology, within a frame within a dualistic polarization of the biological as opposite to the social. Feminist critiques have been haunted by the ghost of biological reductionism and the unhappy and unrealistic binary of nature and culture. The threat of being reduced to the body and defined only or solely on terms of corporeality and anatomy brings together some of the areas of concern of feminism and of studies of sport. Sport can be trivialized because of its associations with play and, especially, within a dualistic framework of a mind/body split because sport seems more concerned with physical than intellectual achievements; in sport too people are reduced to the very source of their success, physical competences, especially when sport is the route taken by those on the margins of society or experiencing social exclusion.

Bodies have been both subjects of theoretical analyses in women's studies and central to political campaigns of the women's movement, for example, which has striven to enable women to reclaim control over our own bodies in aspects of life such as health, reproduction and sexuality. Given that anatomical sex and the particularities of women's embodied role in reproduction have been used to exclude women from full participation in political, social, economic and cultural life, it is not surprising that gaining control of our bodies has been a central political objective of feminism.

Critical theory (Deleuze, 1994; Deleuze and Guattari, 1987; Derrida, 1994; Žižek, Butler and Laclau, 2000) has coexisted in dialogue, and sometimes contention, with feminist scholarship (Beauvoir, 1989; Butler, 1993; Grosz, 1995; Haraway, 1991; Irigaray, 1991, 1989) to develop systematic reflection on socialized bodies. Feminist approaches have, as Young argues, taken as their starting point 'the sociohistorical fact that that women's bodily differences from men have grounded or served as excuses for structural inequalities' (Young, 2005b: 4). Sport is a site where gender differences are created, institutionalized and deeply embedded in the apparatuses of regulation. Although much of the feminist work cited here has not engaged specifically with embodiment in sport, feminist critiques are enormously useful because they present significant challenges to the idea that it is possible to talk about 'the body' without acknowledging and interrogating what makes bodies different and how bodies are treated and valued differently, and often inequitably, and what sort of explanations can be offered for inequalities that are based on bodily differences.

For a long time feminist theorists have seen the project of feminism as intimately connected to the body. They have acknowledged the importance not only of the body as a vital, if contentious, dimension of social relations and of the interrelationship between individuals and the societies they inhabit, but also one in which powerful relations, and those of inequality, are deeply invested which involves recognition that the materiality of different bodies has to be reinstated. Although feminist scholarship has largely not focused upon sport, this concern with bodies is what makes feminist theories so pertinent to the study of sport and especially to embodied sporting practices as reconstituted and experienced through the regulatory bodies of sport.

'The body' or, more specifically, women's embodied experiences have been a major concern of feminist activism as well as being of enormous interest to feminist critical thinkers in women's and gender studies, because the social, political and cultural differences in the ways in which women and men are treated have been attributed, and even justified, by different bodies and anatomical differences. Feminist critiques (Battersby, 1998; Martin, 1998) have shown how the body in Western thought has often been either denied or dismissed, with the mind or the soul occupying positions of 'higher' status than women's embodied experiences, such as those of menstruation, birth, lactation and the menopause. Motherhood has often been an 'absent presence' (Woodward, 1997), both in the systems of representation that constitute culture and in critical analysis (Irigaray in Whitford, 1991); motherhood has been taken for granted. Motherhood's possibility and embodied implications have a troubling presence in sport in terms of regulatory bodies, although perhaps less so for women athletes, especially long-distance runners where postpartum performance can be undiminished and even enhanced. The body is a key site for the experience and inscription of differences, for example, of gender, 'race' and disability and thus has been the focus of analyses of how inequalities are reconstituted through the operation of power; that is of how and why some differences count and the particular meanings they carry. Many feminists, such as Judith Butler in *Gender Trouble* (1990), following Michel Foucault's argument that knowledge is produced, rather than revealed (Foucault, 1973a, 1980, 1988a, 1988b) have been keen to demonstrate that the body *inscribes* rather than *describes* difference; meanings about gender, 'race', 'ethnicity' and disability are produced through the ways in which bodies are inscribed (Butler, 1990). These meanings are not necessarily or exclusively inherent in the bodies which are so differentiated. However, an overemphasis on inscription and social construction might suggest that material bodies do not matter; a problem which has been at the forefront of criticisms of Foucauldian approaches (Martin, 1989; Segal, 1990, 1994) and is addressed in *Bodies that Matter* (1993), where Butler develops the theory of how specific bodies come to matter. Sexual difference emerges from the intersection of the psyche and appearance. Sex itself is created through social practice. This happens at the very moment of naming the infant at delivery; 'it's a girl'. This is what makes the sex of the child and instates a whole set of processes and practices that position that person within the social order of the heterosexual matrix. Although the body is in language, it is never fully of it. Specific bodies come into being through iterative actions and performativity. Through performativity language brings forth that which it names and which is citational, in that language cites practices that are familiar.

Although Butler's work valorizes the body, it de-valorizes gender by conflating materiality and a linguistic understanding of discourse, but this project is achieved with less attention to lived experience than other feminist critiques. However, Butler's work has been particularly innovative and radical in the attempted synthesis of Foucauldian discursive approaches and those of psychoanalysis in a deconstruction of sex as well as gender, which engages with the complexities and intensities of embodied gendered identifications. Her analysis of the processes through which sex itself is made through iterative practices is particularly innovative and challenges some of the most long-standing certainties about both gender difference and embodiment. Butler's critique does distance itself from more sociological accounts or from those that engage directly with women's lived experience and her attempts at demonstrating that bodies matter is problematic, however, when she states that the body is an 'empty sign' that comes to bear 'phantasmastic investments' (1993: 191) that offer an expectation of unity that cannot be sustained. This argument could be seen as compromising the feminist political project of speaking and acting as embodied women. It has been a key concern of feminist theorists, not only to put the body back into theory (and practice), but also to challenge the ways in which the mind/body split has led to the construction of women's bodies as devalued, disruptive and prone to excess.

Theorizing difference

Dealing with difference has been central to feminist theories and many feminists have drawn upon psychoanalytic theory partly because gender difference is also of prime importance in the formations of subjectivity. Postmodernist, post-structuralist theorists like Butler, Grosz and Gatens have worked through different ways of understanding the body as both material and the symbolic and as corporeal and psychic and by exploring the tensions between the imaginary and the anatomical body (Grosz, 1994). What has been called French feminism (Moi, 1987) in the work of Cixous, Kristeva and Irigaray has provided a provocative and interesting route out of Lacanian psychoanalysis and provided a focus on the female body through sexuality, sensation, voice and expression. These critical psychoanalytic thinkers have challenged Lacanian patriarchy and the 'Law of the Father' and the primacy that Lacan accords to the Oedipal stage by pointing to the importance of motherhood and the pre-Oedipal stage of development in the infant's life.

Irigaray's work, although subject to charges of essentialism (Poovey, 1988), presents a powerful challenge to the phallogocentrism of Lacanian psychoanalysis and to Freudian patriarchy, both of which, she argues, prioritize the Oedipal relationship and the Law of the Father at the expense of the mother daughter relationship and the pre-Oedipal imaginary. Irigaray inverts the primacy that Lacan accords the phallus and insists upon the significance of the female body in a feminist project that is based on a politics of difference. Because her notion of sexual difference is both anti-realist and anti-foundationalist refusing concrete reference points that go beyond representation, the symbolic and discursive becomes a privileged site of analysis. Irigaray, like Lacan, notes differences between female and male entry points into the symbolic order and thus into culture, which she attributes to the lack of an imaginary that reflects the mother daughter relationship (Irigaray, 1991). Lack is thus embedded in a cultural tradition and not universal as in the Lacanian version of difference. The phenomenology of bodily experience is constructed through such representation (Grosz, 1990).

Irigaray's focus is upon the ways in which women are symbolized and represented within Western culture, which she argues is based on the death of the mother, whose absence from patriarchal culture, and not on the psychoanalytic assertion that underpins the central Oedipal stage, the death of the father. Irigaray uses material from Western culture that draw heavily upon classical mythology, which elides with Christianity in much of its representations. Even where the mother has a presence as the Virgin Mary the mother of God, it is as an asexual mother who is the vehicle of divine intent (Irigaray, 1991). She points to the 'death of the mother' and the absence of the mother and maternal representations from much of Western culture which is so governed by the father, both as a dominant presence and as a signifier of the patriarchal social order. Matricide is the foundation of the male psychosocial contract as well as of femininity. This provides Irigaray with a point from which to reappraise the maternal as a site for a new, woman-centred genealogies and an alternative symbolic system. Because sexual difference is what forms the basis of women's oppression and the silencing of women's voices and representation in Western culture, she proposes the exposure of the maternal feminine in Western philosophy. Language is vital to this project in order to make the female imaginary and to reinstate the mother daughter relationship within symbolic systems. Irigaray's critique of psychoanalytic thought from within and the attention she devotes to embodied specificity makes her work particularly important in theorizing the body (Irigaray, 1989). Her non-reductionist reading of the embodied self constituted within culture and her positive evaluation of the female body are also particularly relevant in other patriarchal cultural fields. Her focus is on the history of Western philosophy, but the processes she describes resonate across contemporary cultures, none more so than in the field of sport, where her focus upon difference has relevance. A cursory review of current sport reveals a striking volubility of men's sporting achievements and practices and an almost complete silence about women's, apart from particular competitions like the Olympic Games and Paralympics and some sports like tennis and athletics. 'Sport' is still predominantly men's sport, only classified and differentiated by gender when women play, when, for example, it is labelled the 'Women's World Cup' or even 'Ladies' Tennis'.

Feminist critiques are characterized by a focus upon difference, although this has been theorized in diverse ways, have been influential in more recent work, by focusing on the lived experience of gendered bodies, drawing upon phenomenological and social constructionist approaches and putting women's bodies into discourse and by deconstructing previous claims about the gender neutrality of subjectivity which have been seen to be based on men and men's experience. Feminisms had a troubled relationship with bodies in the earlier stages of second-wave feminism because of the dangers that a focus on corporeality might entail the reduction of the self to a biological set of body parts and involve an essentialist reading of gender difference (Woodward, 2008a). Feminist analyses have developed more recently in what has been called the 'corporeal turn' in feminism (Howson, 2005) as part of feminist engagements with the 'problem of bodies' and with rethinking materiality through an attempt to counter some of the apparent disembodiment of some post-structuralist approaches (Butler, 1990, 1993; Grosz, 1994, 1995, 1999). For example, for Butler, everything is discursively created, including sex, gender, desire and body. This generates the problem of the absence of the anatomical body as a relevant factor (Lloyd, 2007). Butler argues that people are sexed at birth and thus 'being sexed and being human are coextensive and simultaneous' (1993: 142). This rethinking of bodies and the intellectual concerns with 'the body' has often been framed by social constructionism, as has been suggested in this chapter. This has had particular resonance for feminist scholars and critics because of the possibilities of political action and social change that it affords. As Jennifer Hargreaves, Patricia Vertinsky and McDonald argue in their collection on *Physical Culture*, 'it is clear that the particular body in question . . . is socially constructed and influenced, changed, adapted, reproduced according to social relations and social structures – and that integral to these processes are unequal relations of power' (2007: 2). By deconstructing the 'natural' body it has been possible to identify the sources and mechanisms of oppression. This is not to say that feminists have been entirely uncritical of the problems that an over-enthusiasm for social constructionism and the more recent corporeal turn have incorporated much fuller analyses of materiality and the complex networks which constitute embodied selves. Particular issues emerge within the corporeal turn, including problems of agency and control and the boundaries of the body and concomitantly of the self.

Sporting bodies

The materiality of enfleshed selves has a special place in sport. Participants in sport are classified by their bodies; almost all sports are differentiated by gender; others include body mass, like boxing. Gendered, racialized differences and body practices in sport have often been attributed to corporeal inequalities, relating to size, anatomy, muscle power and stamina and often elided with psychological aspects of competition. The exclusion of women from many sports has often been based on the claim that women's bodies are smaller and weaker. Being 'weaker' may be translated as being less aggressive, for example, less prone to tackle assertively in football or rugby, or less competitive in contact sports like boxing (Woodward, 2007a).

Exclusion may take the form of invisibility. Women are permitted to participate and engage in embodied sporting practices but do so in unacknowledged ways. Lack of visibility often means lack of endorsement and support which has financial and economic implications. Limited resources and sponsorship mean restricted facilities and financial support for equipment and for training that play or for financing the organizing bodies, including those of the media as represented in the media sport nexus (Andrews, 2006). Sport may appear to be primarily concerned with bodies and body practices, but it is only relatively recently, largely since the 1980s, that theories of embodiment have become central to intellectual interrogations of sport. The corporeality of athletes has mainly been located within scientific explanations of performance and seen as belonging to the realm of biology; that bodies might have social, cultural and even political meanings has not long been part of the repertoire of sport studies. Genetics have been invoked in the attempt to achieve some certainty about the boundaries of sex in sport, which is a field in which regulatory bodies have long sought to fix who can be classified as a woman and who is a man. In practice this has usually been framed by issues of fair play and the claim that men should not be able to gain advantage by passing as women and competing in women's sports. Whatever the fluidity of gender realignment in the wider cultural terrain of liberal democratic societies and the liberal argument that people are the gender they perceive themselves to be, sport has its own rules and measures and focuses on what are understood to be physical, biological and genetic characteristics.

Conclusion

The purpose of this chapter has been to map out the terrain of theorizing the body and embodiment and to indicate some of the major problems that have emerged from the process of rethinking bodies and embodiment. The problems of thinking the body and embodiment have been framed by different tensions and different configurations of the status of corporeality. The socially constructed body is haunted by the limiting frailty and materiality of the flesh creating a tension between biological, anatomical bodies and socially and culturally inscribed bodies. The inscribed body also lacks the energy and physical exuberance of the material body which meets the challenge of sporting embodied practices that extend possibilities. Material bodies are subject to damage, but they are also central to the corporeal pleasure of sporting practices and to their affects. These interconnections between what is social and what is material are reconstituted in the intellectual discussion between the importance of representation and symbolic systems and the experience of lived bodies in phenomenological accounts. The centrality of body projects, which are key to participation in sport, highlights the problem of agency and responsibility in shaping the embodied self that is reflected in the interrelationship between theoretical positions that accommodate self-determination and those that focus upon social, cultural, political apparatuses through which subjects are produced.

The very notion of 'the body' invokes boundaries which postmodernist thinking challenges in diverse ways, demonstrating the impossibility of a bounded self and denying the possibility of a unified subject contained within such a frame.

One of the major problems in developing a theoretical understanding of embodiment is that of providing an adequate explanation which enables the body to 'fight back' without essentializing what it means to be enfleshed. If the body is reinstated, does it mean that we can be reduced to our bodies? The body is part of being and becoming human not its only determinant. The threat of reductionism is one that feminist critics have engaged with in a variety of ways, because arguments based on biology, especially those involving some form of biological reductionism, have constituted a significant element in the genealogy of explanations of gender difference. Some challenges to reductionism go as far as Butler, following Foucault, to argue that as sex too is gendered and normative, similarly bodies too are put into place by that which normalizes them, which reduces anatomical bodies, like sex, to a category with little meaning outside their social context. Other approaches suggest a more positive engagement with the materiality of bodies in challenging essentialism but retaining recognition of the lived body. Reductionist dangers are shared by the marginalized and socially excluded embodied selves who have most strongly resisted them. The political properties of this resistance make feminist theoretical challenges the most productive for the main concerns of this book, although much of the material cited has not specifically engaged with the field of sport.

Some of the intellectual dilemmas outlined in this chapter have particular resonance in sport which is most powerfully defined as *physical* activity as will be demonstrated in Chapter 3 which, by looking at how sport is defined and at what is involved in modern sport, focuses on the embodied sporting practices that make up the participant bodies in sport and their regulatory bodies. The regulatory bodies of sport and the discursive field in which embodied sporting practices are enacted and lived provide a strongly demarcated field in which bodies are very clearly classified and gender categories are strongly defined.

3
Sport: Bodies at Play?

Introduction

Sport presents a distinctive field of enquiry and a particular social world. Bodies may be central to sporting practices, but sport is also characterized by specific ways of acting and ways of being and has its own regulatory frameworks. What is distinctive about embodied sporting practices? What is specific about sport? Sport is a significant part of global and local economic networks like so much of the contemporary economic infrastructure. Sport is a commodity, like many others – a means of making a living for some and of generating wealth. However, sport is a distinctive field, as I shall argue in this chapter, not only because of the centrality of bodies and body practices to sporting projects, but also because of the particular features that mark sport out, including its histories and concern with play as well as with competition and cooperation, and their inherent tension. Sport is also a set of pleasurable activities that people do because they want to. This makes sport attractive, although its associations with play and pleasure may also render it marginal and trivial in some contexts as an area of study and even as a field of endeavour. Sport offers both an equal and a very unequal playing field in which opportunities are both created and denied.

This chapter maps out the field of sport and engages with debates about the significance of sport as play and the relationship between play, games and contests in the historical development of the body practices which make up sport. What is it that distinguishes sport? Sport has evolved through a changing relationship between governance and the apparatuses of regulation in sport and routine body practices. Sports have histories and are always located within the wider narratives of social, political and cultural life. Sporting bodies are temporally and spatially situated within different fields, ranging from specific bodies, such as the International Olympic Committee (IOC), Major League Baseball (MLB), the National Football League (NFL), the Athletics Association (AA), FA (Football Association), National Basketball Association (NBA), International Cricket Association (ICC) and World Boxing Association (WBA), to legislative and quasi-governmental bodies and the media which have become particularly powerful in the recreation and articulation of sport, especially in the media, sport and commercial sponsorship nexus. Regulatory bodies prescribe where the sport can be played and by whom and which body practices are acceptable and which are not. These regulatory bodies in sport determine not only the parameters of what counts as their sport and how they are played, but also some of the relationship between the sport and its spectators. Increasingly, the interior as well as the exterior of the body is the focus of the bodies that regulate and discipline sport. DNA is checked to determine the gender of participants in sport, for example as evident in IOC practices at recent Olympics, and body fluids are assessed to ascertain whether illegal substances like performance-enhancing drugs have been consumed in order to comply with the rules of sports regulatory bodies and the wider sphere of governance.

Sports are also located with specific histories linked to place and culture. In order to explore the interconnections between the embodied selves who participate in sport and the bodies which regulate and shape their sporting practices, this chapter highlights some of the histories of the making of the healthy and 'normal' sporting body and the evolution of regulation in sport, for example in relation to the articulation of gender, sexuality, able-bodiedness and 'race' by addressing some of the literature which engages with definitions of sport. What do we mean by sport? Which practices are included under this umbrella term and which are not? Modern sport is frequently defined by its transformation from what is called ancient or pre-modern sport (Giulianotti, 2005a; Guttmann, 2005; Loy, 1968). Thus, a useful starting point for an exploration of what constitutes sport is to trace some of these ideas about how it has evolved.

From ancient to modern: Sport over time

As Richard Giulianotti notes, sport carries strong connotations of play; we mostly do sport because we want to and not because we have to (2005). Sport is strongly linked to pleasurable activity and to play, which is found in all humans and indeed most animals, at least mammal cultures. Games of physical skill have been found in all societies, past and present (Chick, 2004), although games are not as old as play. In explanations of what constitutes sport, play and notably spontaneous play are seen as aspects of pre-industrial, pre-modern games (Guttmann, 2005). What Guttmann calls 'spontaneous play' is part of the key to understanding the ubiquity of sport/play and its potential. Sport/play, outside the formalities of regulatory bodies, can involve a physical space where identities are not necessarily inscribed or prescribed and what could be a free space. Through its associations with play and corporeality, sport generates creative and productive possibilities that might transcend functional immediate concerns as well as a site for the implementation of regulatory, disciplinary practices.

The sort of body practices which are classified as sport are at least as old as the first recorded Olympic Games in Greece in 776 BCE, although the modern phenomenon of sport involving some sort of coherent organization is not found in all past societies and certainly not everyone was permitted to play, or even to watch sport. Notably women were excluded and were only ever able to attempt to watch if disguised as men, at considerable risk to their lives. Sport has long been governed by aspects of visibility, which marks gender difference as an example of noting not only who is allowed to participate, but also who is allowed to watch. Clothes have been central to visibility and display in sport and to the regulation of sport. In the ancient games athletes largely competed without any clothes, but contemporary participants wear body revealing attire that has gendered implications. Wearing appropriate clothing that involves the freedom to display their bodies in sporting practices in the public arena remains problematic especially within some cultural, religious contexts in the twenty-first century, although such restrictions make sport a site of resistance where such rules can be challenged (Fozooni, 2008). Many contemporary sport forms have long cultural traditions and historical legacies. Sprinting, long jump, javelin and wrestling were all part of the ancient games. However, it is evident that modern sports have diverse historical roots and social derivations and there are different classificatory processes in play.

The corporeality of modern sport reiterates body practices that can be construed as basic body movements such as climbing, diving, kicking, jumping, running, swimming, throwing, vaulting and weightlifting (Ingham and Loy, 1973). Many events of the modern Olympic Games are based on such basic body practices. Some pre-modern as well as modern work as well as play and practices such as fishing, hunting, skating, sledging and skiing have also persisted as sporting practices. Many such activities involve animals, such as dog racing, horse racing, pigeon racing and rodeo events which may also be assigned to this category, as they represent the transformation of human use of domestic animals for work to purposes of play. Some activities that are commonly called sports involve killing animals, as in field sports. The transition to modernity, however, seems to involve classification of an instrumental, functional activity, such as killing game, which was for consumption, into a version of sport as play. By the twenty-first century, birds are raised solely for sport, overfed and kept largely immobile so that there is little challenge in hitting them in flight and the quantities shot are so large that most of the birds are discarded rather than eaten. In cases such as these there is limited pleasure either in the sport itself or in the consumption of the birds so shot. Field sports, such as hunting with dogs, have become political issues in recent years, especially with increased state regulation. What is classified as sport is the site of often acrimonious conflict between different positions; one side emphasizes the cruelty of the practices and the other stresses tradition and the rural economy that is often partly sustained by such sport.

Pre-modern body practices that have become sports also include those linked to pugilism and warfare including ancient martial arts and military exercises. Sports such as archery, boxing, fencing, javelin throwing and wrestling are immediately recognizable examples. Some of these sports retain their militaristic associations into late modernity and many, if not most, men's sports are frequently described in the language of pugilism. A point not so readily made by commentators is the elision between bellicose body practices often linked to militarism and the associations of aggression and competition that also accompany particular versions of masculinity in sport. Whilst some of these sports have become regulated and refined through new technologies which restrict risks, others, notably boxing, are more contentious. There is a quite vociferous lobby which argues that pugilism has no place in sport.

Another dimension of traditional sporting body practices is that linking them to festivals and fairs; a number of modern sports have distant roots in ball games, dances and ceremonies associated with the religious practices of traditional, pre-literate societies. Many modern sports developed out of pre-modern and ancient games enacted at pagan rituals and medieval fairs and festivals. Although modern sports may differ markedly from their original folk forms they nevertheless possess significant residual sporting traditions, styles and practices (Ingham and Loy, 1993). Sport retains aspects of the carnivalesque (Bakhtin, 1984; Stallybrass and White, 1986), although probably more in the rituals of spectatorship than in the embodied practices on the pitch (Bale, 2000; Giulianotti, 1991, 1995). Fans may dress in grotesque or bizarre costumes and exaggerate their gestures, which are framed by the ethos of carnival, but deportment on the field is unlikely to be as invocative of excess in quite this form. Some sports have been reconfigured to incorporate some of the carnivalesque into their performance. Boxing goes much further than card girls with music and some dramatic ring entries; cricket, since the advent of Twenty20 and the IPL, offers spectacles and drama on the field as well as off.

An argument which has strong support in the literature is the claim that modern sports represent a continuum of development from informal and often brutal play to formal competitive play, through athletic folk games, to recreational and representational sports (Elias and Dunning, 1986). This can be traced through the regulation of particular embodied sporting practices. For example, it has been argued that forms of folk football led to the development of soccer (Chick, 2004), which led in turn to the development of rugby (Dunning and Sheard, 1979) and in turn to the development of US and Canadian gridiron football (Riesman and Denny, 1951). This demonstrates some aspects of a linear narrative of progression, although these are a few examples and not the whole story. Sporting practices are constantly transforming along an uneven and often disrupted trajectory.

Historical dimensions: Politics, places and practices

Although some commentators use the device of a linear narrative of change, the complexity of the story is largely unacknowledged. Modern sports are seen to have emerged through a variety of historical and cross-cultural influences, which reveal both the centrality of power relations in sports' social history and important cultural differences in the foundation of modern sporting traditions and institutions (Giulianotti, 2005a). Developments and transformations of sport, from play and games, have also been shaped by articulations of the politics of gender, class, race, ethnicity and disability. The bodies of those who participate or of those who have been permitted to participate constitute the political activism which has created new sporting regulatory bodies and reconfigured the relationship between body practices and organizing bodies. This has happened with the development of women's sports organizations, the disability movement including the Paralympics and black and ethnic minority and anti-racist organizations.

Social class has always played a significant part in the development of sport, along with colonialism. The upper classes in the US had enjoyed tennis, polo and cricket, but North America was much less influenced by the games of the British Empire and forged their own distinctive national traditions at the start of the twentieth century. Global superpowers have a tendency to lay claim to having 'invented' modern sport (Smith, 2008). It is likely that most of what is now classified as modern sport has arisen from sets of body practices that either constituted play or training for pugilism or catching food to eat, but it is the rules that usually do have links to place. Baseball may have developed in Europe (Block, 2005). It has recently been claimed to have been devised in Britain and mentioned in Jane Austen's late eighteenth-century novel *Northanger Abbey* (Norridge, 2008). Such claims have been greeted with some surprise, at least in part because a game that had been spread initially by the army in the middle of the previous century and gained mass popularity among the working classes in the twentieth century in the US has such powerful associations with traditional, working-class masculinity. The recent debate might have been expressed in terms of place and the US or the British origins of the sport, but gender is what is probably most troubling. A sport that could be played by genteel young ladies in floor length dresses might resonate with rounders – a sport often played by British school girls, but is unlikely to be a welcome moment in the history of a sport that is so firmly imbricated with working-class masculinity. Young men at leading universities took up American football; meanwhile, the Christian movement invented basketball and volleyball as alternatives to existent sports. Canada's national sports were ice hockey and lacrosse, the latter developed from the game played by the indigenous peoples.

Some contemporary North American sports are reputedly the result of individual invention and their histories are told within the US narrative of heroic individualism that is characteristic of the construction of hegemonic masculinity. Classic examples are basketball, which was invented by Canadian James Naismith in December 1891 while a student at the YMCA Training College in Springfield, Massachusetts, and volleyball, which was claimed to be invented by William G. Morgan in 1895 while serving as physical education director at the YMCA in Holyoke, Massachusetts (Ingham and Loy, 1993). These are typical of the narratives of sporting history and the literature of the development of modern sport. Although some stories, such as the invention of over arm bowling in cricket by women in long skirts (Major, 2007; Williamson, 2009), invoke novelty rather than heroism or a substantive myth of origin that has status within the annals of sport. Such attributions do reinforce the traditions of sport and the synergies between gender, class and embodied practices.

Sports are taken up in different places, for different reasons. Latin America has one of the strongest soccer traditions. Football is a way of life and a big part of everyday popular culture although baseball is hugely popular in Central America. In Japan, traditional martial arts were transformed into sports like judo and karate and achieved wider participation with some translation according to Western criteria too. There has been some two-way traffic in sport. Thai boxing has also spread to Europe and the US, for example, with a range of interpretations and schools many of which constitute a holistic approach which integrates sporting body practices with world views and ways of living. Baseball, in modern Japan, has long-standing popularity; reputedly dating from the famous all star tour of 1934 which featured Babe Ruth, then in his last year with the New York Yankees, the first American superstar whose own fame might even supersede that of his sport. Gymnastics remains important in school curricula, but football is now rooted within Japanese popular culture (Giulianotti, 2005c).

The British, and more specifically the British Empire, played a particularly important role in transforming games and pastimes into codified sports, and then transmitting these cultural practices across the globe. Until well into the nineteenth century, many sporting practices had centred upon horse racing, blood sports and prize fighting and associated gambling and betting practices (Mangan, 1998a), but discipline was enforced and spontaneous, unruly energies contained through the establishment of sports like football, rugby, field hockey, boxing, lawn tennis, squash, and track and field athletics (Mangan, 1981). The regulation of sporting practices and creation of rule-governed sports could be seen to counter the excesses and licence of the drinking and gambling linked to previous games, although an alternative view is to see the licence of drinking and especially unregulated gambling as key components of the genealogy of sporting practices, which persists into the twenty-first century, albeit in different and more clandestine and global forms. Sporting regulations and regulatory practices operate at different levels, temporally and spatially. Gambling is a routine practice and ubiquitous in many sports, but though prevalent its extent is largely not explicit. Gambling opportunities still dominate the advertising in the sports press and on the home pages of the Web sites of major sports, for example the English Premiership club home pages. As I demonstrate in this book, some of these aspects of pre-modern sport remain associated with the gendered networks of sport and are interconnected with particular versions of sporting masculinity. Masculinity and gendered differentiations are routinely imbricated in the embodied practices of sport and through its regulatory bodies.

British influence overseas opened up distinctive channels of international diffusion, along imperial and trade routes, for different games. This sporting influence was also strongly class-based, expressed by some commentators in a class-inflected language of colonialism: 'where the public-school boys went in large numbers, inside or outside the Empire, there cricket and rugby prevailed, and where the horny-handed sons of toil, or at least the counting house, predominated, there soccer fever tended to infect the locals and become endemic' (Perkin, 1989: 117). Sport became powerfully linked to a gendered culture in Australia and South Africa, countries which also contributed to the codification of sporting practices. The Australians were the first to regulate football as their sport with Australian Rules in 1859. Australians have excelled in most sports, particularly in cricket, rugby, league and union and swimming. Across Southern Africa, football and boxing became particularly popular. Football, boxing and athletics remain the top sports in Africa, although obviously with variations across countries. Footballers from Western African countries like Nigeria and Cameroon have recently achieved considerable success in European football leagues, whereas Eastern African countries have produced particularly successful runners. The success of football in the continent is evident in the South Africa staging the Men's World Cup 2010; the first time the competition has been held in Africa. In the Indian subcontinent, cricket was favoured, although local peoples challenged and reinvented its intensely colonial value system. After national independence in 1947, different local rules were harmonized and playing procedures standarized (Giulianotti, 2005c). More recently, the Indian Premier League (IPL, 2008) has transformed cricket through the development of the short game of Twenty20 and Day–Night games as popular mass entertainment performed by high-earning players. This is a contemporary example of the dynamic between the regulatory body, in this case the IPL, and embodied sporting practices on the field where players take risks with big hits, pull shots, hard drives to score quick runs and running between the wickets in order to up the scoring rate in the short game. However, there are also specificities of culture and place that impact upon how a game is played; cricket is most powerfully enmeshed with other aspects of life in India and for many current stars of test cricket, like with its complex, lengthy matches over 5 days, see Twenty20, as an extension of their beloved game and another version, rather than a threat to its authenticity as Sachin Tendulkar argues (Brearley, 2009). There is evidence of both the development of imported games and sports along local lines and of local games that become codified and regulated with the advent of modernity. The Beijing Olympics in 2008 drew the world's attention to the excellence of Chinese athletes as well as to different understandings of histories of sporting body practices (Spence, 2008).

Other sports retain memories of a colonial past, even though most sports, not just cricket, have been reconstituted in post-colonial times. Tennis still has links to the Queen's Club and Wimbledon, even though the major tournaments, which dominate the sport, are in the US, France and Australia and the top players in the twenty-first century are drawn from wide geographical locations.

Different European countries have played by different rules. In Germany in the early nineteenth century, the *Turnvenein*, a set of nationalistic disciplines combining gymnastic drill with military training, in particular, fencing, was founded by Friedrich Jahn (Segel, 1999: 209–211). The Czech equivalent, *Sokol*, spread across Central and Eastern Europe. Handball was organized and popularized in Germany after the First World War to challenge football's popularity. France's major contribution has often been political and administrative in the development of regulatory bodies, promoting a remarkably prescient, cosmopolitan and international vision of sport's global potential. Baron Pierre de Coubertin founded the modern Olympic Games, first contested in Athens in 1896. The French provided the impetus for the establishment of football's governing body, the International Federation of Association Football, FIFA (Fédération Internationale de Football Association), in 1904. France's distinctive sporting event is the Tour de France, the world's most prestigious cycling race (Thompson, 2006). While football's European hegemony remains intense, spatial, and cultural and climate circumstances mean that shooting and the alpine sports are strong across the Alps and Nordic nations. Other sports have developed as part of the assertion of local cultural identities. In Ireland, hurling and Gaelic football were established to counteract perceived British cultural imperialism, and have maintained strong grass-roots support.

These shifts over time, ranging from rural, primitive athletic folk games of old, to new urban, hi-tech, extreme sports and spatial variations, demonstrate some of the diversity of the body practices that constitute sport. A single definition of sport is unlikely to fit so varied a range of sporting practices. This brief outline does, however, highlight the major social characteristics of modern sport and suggest the specification of the embodied material properties and social processes underlying the social development of modern sport. It also provides a set of common features for examining the magnitude and complexity of sport as a material and social phenomenon at different levels of analysis, including sport as a unique game occurrence, sport as a particular type of ludic activity, sport as an institutionalized game, sport as a social institution, and sport as a form of social involvement. These social dimensions of sport open up what sport shares with other fields as well as signposting some of sport's specificities.

Sporting definitions: What's included?

What distinguishes sporting practices and sport? What constitutes sport can be seen to have some generic characteristics, usually involving aspects of being an embodied, structured, goal-oriented, competitive, contest-based, ludic, physical activity. As Richard Giulianotti points out, definitions of the word 'sport' are slippery, and carry significant pre-industrial associations with aristocratic field activities like hunting and shooting and a very class-based set of sporting pleasures. He lists dictionary definitions which encompass the verb to 'sport' as to 'frolic', 'make merry' or 'amuse' as well as to 'wear' and 'exhibit'. The noun 'sport' denotes 'recreation', 'games' and 'play'; or 'amorous behaviour', 'mirth', 'jest' and 'dalliance' (Giulianotti, 2005a). He suggests that the social actor, the 'sportsman', is plainly a 'good fellow', which also firmly locates the gender of those who engage in sporting practices. The 'sportsman' is also a 'gentleman' who is, by implication, honourable and knows the social as well as the sporting rules. The 'lady' or even more so 'woman' would have very different connotations. Women's sport has often used the nomenclature of 'ladies' to distance participants in sport from more working-class engagements in physical activity. It is only relatively recently that the description of sport as 'women's' has become almost universal. Sports wear and equipment are still advertised as for 'ladies' (e.g., Nike and Adidas advertising and Web sites), whereas 'gentlemen' is largely reserved for hunting and fishing or historical accounts. The British public school tradition was to accord girls' sport a class-based status, reflecting the tradition in boys' schools, but with a much less central role, especially in terms of corporeal engagement. The gendered aspects of sport which are often taken for granted and unstated in explicitly differentiated language, when men are involved, are powerfully linked to definitions of modern sport in much of the literature. Definitions of sport, especially modern sport greatly stress its competitive, goal-oriented, contest-based qualities, which the inclusion of its ludic, playful dimensions seems to contradict. However, there is also some elision of the contests and militaristic associations of pre-modern sport and the competition of postmodern sport.

Sport involves bodies and rules. McPherson, Curtis and Loy (1989) suggest that sport is always governed by rules and codes of conduct, which are spatial, such as courts, pitches and playing fields, and temporal, including time limits on the games and the frameworks and institutions of government. Sport is seen to be goal-orientated because its practices are aimed at particular objectives, such as winning points, scoring goals, winning contests or increasing averages and thus winners and losers are identifiable. Sport is also largely competitive in that rivals are defeated and records are broken. The ludic elements of sport relate to its pleasurable dimensions, because sport enables playful experiences and generates excitement. Sport is also culturally situated in that its characteristics correspond closely to the value systems and power relations within the society that hosts the sport (McPherson et al., 1989: 15–17).

Not all physical activities are sport, of course; body practices, even when performed in the open air, do not necessarily constitute sport. Sport is usually distinguished from other body practices like walking or exercising that lack competition. Some walking does, of course, follow strict rules and it does involve an element of competition, for example fell walking. A focus upon competition is, however, problematic, since increasingly activities like walking and working out in the gym have become central to maintaining physical fitness and well-being. The narrow definition of sport in what can still usefully be called 'malestream' (Daly, 1978) thought, as necessitating competition, could be seen as excluding those who seek the physical and personal challenges of embodied practices without wanting to defeat opponents. Despite their competitors' more limited physical activity or fitness levels, games like darts, bowling, snooker, pool and motor-racing are usually included within the category of sport. Each requires intensive physical engagement and proficiency in hand–eye coordination, which also entails a level of physical fitness, but not at the level required for the most physically demanding sports like the major team games, athletics, tennis or squash. Each sport is structured, goal-orientated, competitive and ludic. There is some disagreement about what constitutes a sport as is apparent in the literature which engages with categorizations (e.g., Guttmann, 2005; Ingham and Loy, 1993; Loy, 1968), but all involve the two elements with which this book is concerned, the body practices of embodied selves and a regulatory framework of some sort, which could include the regulating techniques deployed by those who engage in embodied sporting practices.

Allen Guttmann provides one of the most authoritative accounts of sport with a list of characteristics that can be broadly classified within the two interrelated versions of bodies which are the focus of this book. He claims that sports are universal phenomena, but that each human society has its own definition. He cites the enormous diversity of games and pastimes, including board games, card games, dancing, cycling to work, window-shopping and sunbathing, apparently included in UK's General Household Survey (in Guttmann, 2005). The inclusion of such activities demonstrates how boundaries are being crossed between leisure and sport with a shift from a bounded categorization of sport per se. Guttmann's definition of sports sites them within a paradigm that is quite specific about the relationship between play, games and contests. Sport, however, has particular features. It is structured with rules and regulations, it is competitive with goals, it retains some notions of play and enjoyment in what the sports sociologists call its ludic dimensions and it is physical.

Structures and regulations: Playing by the rules

Modern sport is highly structured. First, sports are structured in some ways whether they are played formally or informally; a knock about on the tennis court or a kick around in the park or in the street with a make shift ball and coats for goal posts still follows rules. All sports are rule-governed by either written or unwritten rules; these rules are understood, however rudimentarily, by all who participate. Rules may be embedded in the body practices which make up the sport so that they become taken for granted or, at the other end of the spectrum, they can be highly formalized and complex, requiring considerable expertise to grasp the full range of regulations. Second, most sports are temporally circumscribed as illustrated by designated time periods such as innings, halves and quarters, or number and time of bouts and rounds, or allocated attempts within a specific time period. Time scales have been adapted in particular instances to prevent indefinitely long sporting encounters and sports have instituted tie-breakers, sudden death playoffs, and shorter versions of selected sports, for example in one-day cricket matches and more recently the development of Twenty20, Day–Night games. Women often participate for shorter timescales and cover less distance or play fewer sets or games. Third, most sports are spatially circumscribed by the sites of their venues, whether these are arenas, courts, fields, pools, rings, rinks, stadiums or tracks, with the most privileged, elite athletes accessing the most superior sites. Lastly, modern sports tend to be formally administered, whether by local clubs, universities, colleges or schools, professional teams, or sport federations which can be regional, national or international.

Goals and competitions

Most of those who write about sport argue that sport has to be competitive (Cashmore, 2005; Giulianotti, 2005a; Guttmann, 2005; Houlihan, 2008; Loy, 1968; McPherson et al., 1989; Tomlinson, 2007). Competition may be between individuals or teams, and may involve either an animate object of nature such as a bull in a bullfight or an inanimate object of nature such as attempting to compete against nature, for example in climbing the highest mountain in the world, or it may be focused on competition against an ideal standard (Loy, 1968). Competition in sport is allied to its goal-oriented characteristics, which are stressed by all commentators who write about the more generalized field of sport (Giulianotti, 2005c; Guttmann, 2005; Loy, 1968). Ethnographic accounts are more likely to accommodate the personal satisfaction which athletes gain from their physical engagement (Messner and Sabo, 2000; Sugden, 1996; Wacquant, 2004), especially those which address the gendering of sport and include women's participation (Hargreaves, 1994; Hargreaves, Vertinsky and McDonald, 2007; Robinson, 2008; Wheaton, 2004). More recently, there has been greater emphasis on the personal pleasures of sport, especially in non-mainstream sport, such as skateboarding, windsurfing, mountaineering and fell running, and the development of the notion of post-sport (Atkinson, 2008; Robinson, 2008; Wheaton, 2004). In accounts which seek to offer a definition of sport (Giulianotti, 2005a, 2005c; Guttmann, 2005; Loy, 1968; Ingham and Loy, 1993) individuals, teams and organizations are seen as typically goal-directed in sport situations, especially in terms of the overriding goal of winning. Athletes and coaches continually attempt to achieve various

standards of excellence that are associated with the highest levels of achievement in the hierarchical taxonomies of their sport.
Numerous forms of self-testing and con-testing take place in all sporting encounters. The sporting media constantly stress the theme of
being top in terms of games won, points earned, medals obtained,
rank on the money list, most career victories or number of Grand
Slam titles.

It is argued that spectators are drawn to competition (McPherson
et al., 1989). This is most dramatic in a sport like boxing, where there
are only two competitors in the ring (Oates, 1987), although team
games attract the most passionate of affiliations as, for example, in
baseball, basketball, cricket, football and soccer. There is also parallel
competition whereby participants compete against one another indi-
rectly by taking turns as in bowling or golf; or contesting in separate
spaces as, for example, separate lanes in swimming events or track
sprints and hurdle races. There are also forms of competition which
are largely competition against a standard such as trying to make a
qualifying time for an Olympic running event, or attempting to set
a world automobile speed record such as the famous attempts on the
Bonneville Salt Flats in Utah, in the US.

Sport is ludic: All this is supposed to be fun?

In Guttmann's paradigm of the move from play to games to sport he
acknowledges the persistence of the playful elements as well as the
mutability of meanings about sport, even going as far as to suggest
that sport might mean what those who participate claim it to mean;
sport is made in its spatial and temporal context. However, a persist-
ing component of the body practices which make up sport is that
even the most highly professionalized forms of sport possess some
play-like elements.

There is no single reason for sport's huge cross-cultural appeal.
Sport does appear to be a kind of human medium that joins people.
Modern sport promises playful pleasures to players and spectators;
new skills are tutored and learnt and the experiences can be enormously pleasurable in themselves, without necessarily having outcomes measurable by the criterion of objectively assessed success.
Different sports facilitate controlled, pleasurable interaction with
particular landscapes. In our increasingly performative consumer culture, the physical endeavour of sport compensates for sedentary

working practices and provides the exhilaration of total immersion in corporeality.

Physically competitive play, agonal games, and elite, professional sports are similar in that they usually involve competition between two or more sides, with agreed rules, criteria for determining the winner, and the outcome largely based on the display of superior physical skill. Because they share the same basic features, play, games and sport are often treated as one and the same. These ludic activities are attractive and appealing for participants and spectators alike, at least partially because they are fun (Huizinga, 1955) which is why through their sociability, euphoric interaction, quest for excitement and emotional dialectics sport is fun too. 'Fun' and even 'ludic' might seem to trivialize the affects of sport. 'Fun' could underestimate the deep personal and collective emotions experienced by participants and spectators. Sport engages a whole range of emotion, affects and corporeal, sentient and sensual pleasures.

Physicality: Bodies in sport

Much of the debate about definitions focuses upon what sport means or signifies. Sport may present a discrete area of experience, characterized by its own sets of rules and practices, but many accounts locate it within the context of what it means, thus laying great stress on the signifying processes through which meanings about sport are reproduced. Guttmann cites the case of anthropologists who emphasize a more local definition, claiming that sport means what people want it to mean (Grupe, 2005 in Guttmann, 2005). Such definitions also address the pleasures of sport in more meaningful ways. Alternatives invoke the notion of essences that imply something more fixed, as does Bernard Jeu, who suggested that all sports were a meditation on death and violence (Jeu, 1972) in an elevated version of the more prosaic claims of Bill Shankly the then English Premiership football club, Liverpool football club, manager, whose apocryphal statement that he was disappointed with the claim that football was a matter of life and death, because it is much, 'much more important than that', sums up the more grandiose claims for sport's essential and profound qualities (in Goldblatt, 2006). What is implied by the somewhat unlikely congruence of thought between an academic's reflections on the profound essence of sport and the hyperbole of a practitioner in the field of popular sport is sport's very physicality and the feeling that it must be about more than signification and representation. Sport is material and social in its inextricable interconnections.

The degree of physicality varies by sport, but the body constitutes both the symbol and the core of all participation in sport (Hargreaves, 1994). Embodiment in sport manifests in the ways in which sporting activities involve many kinds and degrees of physicality, including physical activity, physical aggression, physical combat, physical exercise, physical presence, physical prowess, physical recreation, physical sexuality, physical training and physical work. Sporting bodies represent a range of desiring bodies, disciplined bodies, displaying bodies and dominating bodies, which means that physical bodies are pivotal to any discussion what constitutes sport. The range of embodied practices covered by sport is also central to an analysis of selfhood. As Brian Turner writes,

> The appropriateness of bodiliness in all its aspects, from sexuality and reproductive capacities to sensory powers and physical health, strength and appearance, is the fundamental matrix, the material superstructure, so to speak, of the production of personhood and social identity.
>
> (Turner, 1994: 28)

Embodied selves are recreated in space and time and sport offers one of the fields in which this takes place with a self-conscious focus upon corporeality which is not always present in critiques of identity and the making and remaking of selves. The incorporation of physicality into all the definitions of sport provides a useful site for exploring the importance of the body.

Explanations of change: Pre-modern, modern, postmodern sport

It has been argued (Guttmann, 1978, 2005; Giulianotti, 2005c; Loy and Ingham, 1993) that the formal-structural characteristics of modern sports are significantly different from those of pre-modern sports. Pre-modern sports have either had to change to acquire the formal-structural characteristics of modern sports, or to survive on the margins of the mainstream. Guttmann suggests that evolution from pre-modern to modern sports is an instance of what Norbert Elias called 'the civilizing process' (1978) through which the members of a society internalize values that reduce the level of expressive interpersonal violence. *The Civilizing Process: A History of Manners* (Elias, 1978) describes the developments by which manners, standards of conduct and the body practices of social behaviour were transformed from the sixteenth century onwards, becoming reconstituted in the eighteenth and nineteenth centuries into acceptable, respectable codes of practice which limited violence and incontinent manifestations of excess. Individuals exercised greater degrees of physical and emotional control and constraint. Elias's explanation offers an empirically evidenced trajectory of change which has the considerable advantage of focusing upon embodiment and corporeality in relation to the impact of regulating practices and regulatory bodies. It is, however, a somewhat linear narrative and one which has more difficulties in accommodating disruptions to the storyline of civilizing processes.

Interrelated processes which link the self-regulation of individuals to social trends which include economic growth, sophisticated patterns of production and consumption and a complex division of labour have all led to what Elias called the civilizing processes which has implications for the move from pre-modern to modern sport with the imposition of rules that limit violence and spontaneity. Although Elias accepted that sport can heighten emotion, it increasingly serves a mimetic function as an outlet for emotions that social norms require to be controlled. This approach resonates with the popular belief, shared by policy-makers historically from at least the late nineteenth century, that sport can channel emotions that would otherwise be repressed or erupt inappropriately, but gives less weight to the idea that sport could be productive of such emotions and feelings, either on the field or the court or in the ring in creating aggression and an excess of competitive spirit or in crowd behaviour that crosses acceptable boundaries.

Explanation is closely linked to definition and to transformation in the key elements which make up sport. Guttmann draws on a Weberian analysis of bureaucracy and rationalization applied to sport, combined with Elias's understanding of the processes involved in the progress of civilization, which has the advantage of stressing the impact and influence of organizational structures in shaping modern sport. Elias's notion of the civilizing process resonates with

Weber's understanding of rationalization (1978) which represents a process of systematization and standardization whereby social practices, like sport, are subjected to abstract, explicit, often impersonal rules and procedures which shape what can and cannot be done as well as what is included within this set of practices, in this case what counts as sport.

Guttmann claims that modern sports can be defined by a set of seven interrelated formal-structural characteristics (Guttmann, 2005: 24–28), which constitute their regulation. Thus, modern sports are characterized by their secularism and are not related to the transcendent realm of the sacred, which was the case for sports in pre-modern societies. The Olympic Games, for example, were sacred to Zeus, and Japanese *shinji-zumô* was performed at temple and shrine festivals. This is a view that has been challenged given the extremes to which contemporary sports fans will go in supporting their team but Guttmann's point concerns the formalization of sport rather than the practice.

Guttmann's second characteristic of modern sport is equally contentious in some ways although it clearly has some purchase. He argues that modern sport is based on principles of equality; rules must be the same for everyone and everyone should be included. Whilst the first aspect of this principle concerning rules on the field, track or in the ring clearly holds, the openness of modern sport is subject to question. Pre-modern sports featured more explicit, formal exclusions than modern sport; women were barred from most sport including the ancient Olympics whereas the modern games offer almost equal numeric participation with few actual sports from which women are excluded, although women do compete according to different rules in most sports, for example, in terms of times and distances. At different times and in particular sports gender differentiated rules demonstrate that this is not a simple trajectory. For example, different sets of rules can be designed to promote equality by protecting and acknowledging embodied difference. In such cases there are separate single-sex competitions for women and for men; boxing, weightlifting, wrestling have largely prohibited mixed competition, although this is disputed, for example, at some weights in boxing. In many situations banning mixed competition might be seen as promoting gender inequalities, where women, for example, run shorter distances, hit of shorter tees; it is claimed that the best women athletes can only improve by competing with men, as has been argued in athletics and in golf, where both Annika Sörenstam and Michelle Wie attempted, at different times to enter men's competitions. Sörenstam in particular was subject to disparagement from male golfers like V.J. Singh (2009) and from the media critics for her attempts to compete with men, who would at the time include greats such as Tiger Woods. Her treatment by the news and sports media suggests the rules about who can and who cannot participate extends beyond the rule book and the media play a big part in creating patterns of exclusion Although neither Sörenstam or Wie succeeded and Wie failed to make the cut on several occasions in her first year as a professional in 2006, their attempts raise the question about how rules apply and challenge both the universalism of equality and the notion of a simple path of transition between pre-modern and modern sport. In this case, 'rules' could be too narrow a concept to describe the set of regulatory practices that reproduce and circumscribe embodied practices in sport. Modern sports are not only rule bound in complex and uneven ways, but also deeply implicated in the perpetuation of some inequalities, many of which are significantly embodied. Guttmann acknowledges that men have always decided

> which physical contests are socially and culturally important and which are not. I am ready to welcome a history of sports that pays as much attention to women's contests-of all sorts-as it does to men's contests, but the source materials for this kind of radically revisionist history are not yet and may never be, available. (2005: 25)

The unlikelihood expressed here about women's sport ever being 'put into discourse' demonstrates not only the power of classificatory, regulatory systems in defining what constitutes the legitimacy of the field, but also the gendered specificities of this in sport.

Modern sport involves greater specialization, the third feature in Guttmann's discussion, than its pre-modern counterpart. Modern sports such as American football, rugby and soccer have developed from earlier games and modern games are marked by a high degree of detailed specialization in their rules and customs. The practice of sport has become regulated by medicine, science, technology and psychology. All elite athletes, women and men, are now accompanied by teams of experts who represent specialist knowledge, although global inequalities and the unequal distribution of resources persist; for example, many elite athletes who started their careers in Africa or the Caribbean have only been able to access these specialist teams by transferring to the US.

A further, strongly Weberian feature of modern sport identified by Guttmann is that of the bureaucratization that characterizes almost all sport, from local, community clubs to international bodies like the IOC, which itself has national bodies (National Olympic Committees, NOCs). Children's sport is similarly bureaucratized, for example, with Little Leagues, junior competitions and under 11 teams. This bureaucratization is powerfully linked to the rationalization, which constitutes another of Guttmann's seven features of modern sport. Sporting competitions take place at purpose built venues, where scientifically disciplined athletes compete for specific rewards according to highly regulated systems of assessment, which make up the rationalization which Guttmann also identifies as one of the distinguishing aspects of modern sport, which marks it out from pre-modern sporting activities or games. For example, the ancient Greeks had much more rudimentary ideas about the timing of contests and distances; the unregulated, no-holds-barred one-to-one fighting, *pankration*, a fore-runner of bare knuckle fighting, most likely ended with the death or total incapacity of one of the combatants, rather than the carefully timed bouts of modern boxing, which is not to say that modern boxing does not, if rarely, end in severe damage to participants or even fatalities. Athletes competed against each other without the mechanical timing, which has reached heights of sophistication in modern times with records being broken by fractions of a second.

The obsession with records is the last of Guttmann's characteristics of modern sport which is clearly linked to technologies of measurement and rationalization. We can be obsessed with records because we have the means to measure them. A sports record is an unsurpassed quantified achievement and apparently a constant challenge to those who seek to surpass it. Other criteria of evaluation could have applied to pre-modern athleticism, including aesthetic values, which occasionally feature in contemporary sports writing but quantifiable, objective measurement is what secures records and is primarily the yardstick of success. The emphasis on measurement records and record-breaking extends beyond the technologies which facilitate and make possible such obsessions; it is also underpinned by the intersection of a power geometry that favours those who break the records and frames the parameters of what constitutes success. Precise technologies of measurement create new criteria for success and increase competition by providing greatly elaborated and enhanced possibilities of evaluation and a whole new science of measurement. Guttmann's characteristics are interconnected and each is implicated in the power relations in play within the wider social and cultural terrain in which sports are played. What is lacking is the focus on difference and especially inequalities which informs feminist and post-colonial critiques. However, Guttmann's arguments remain useful as a measure of the transformation from pre-modern to modern sport.

Guttmann does, of course, acknowledge the limitations of his classification of the features of modern sport; some ancient sport was measured, records were kept even not quite as precisely as modern day timekeeping technologies permit. The picture is much more complex than the starkness of his initial taxonomy admits. He accepts that the modernization theory on which these features are based could be seen as a facile instrument of ethnocentric, ethical evaluation, especially in its more optimistic associations of the progressive march of civilizing sporting practices (Guttmann, 2005). There are clearly applications for this paradigm and it remains pertinent, especially highlighting some of the key features of sport and in understanding the place of sport in the contemporary world. Some of the complexities and disruptions in the narratives of change are addressed in discussion of what could be called postmodern sport, although postmodern should not be taken as part of a simple chronology of change following anything like a linear path.

Postmodern sport: Sportification and post-sport

There are different ways of looking at contemporary sport, the place that embodied sporting practices occupy in the world in the twenty-first century, and, linked to this, how contemporary sport might be defined and categorized. Much of the usefulness of classifications of sport relates more to the gaps between pre-modern, modern and postmodern sport than to the ethical framework within which it is positioned. Modern sport has been significantly transformed

by globalization and its links to what has been called 'sportification' (Ingham, 2004; Von der Lippe, 1994). Modern sport cannot be
disassociated from the global financial networks of commerce and
sponsorship through which it is constituted.

In what are called modern times in the literature of sport, sports
have been transformed from primarily expressive activities to largely
instrumental activities through what has been called a process of
sportification (Renson, 1998). However, postmodern sport, or what
is sometimes called post-sport (Rinehart and Sydnor, 2003), might
be seen to be recuperating some of the expressive and affective
aspects of pre-modern sport, albeit in very different ways. International sport has changed from contests between individuals and/or
teams to contests between nation states that have unequal resources
to produce elite athletes and teams. The twentieth century saw the
articulation of professionalism and Olympism as the two dominant
sport ideologies merging into a single organic hegemony (Donnelly,
1996). Donnelly argues that this trend has been self-reinforcing in
that it marginalizes alternatives and becomes a standard against
which other forms of physical culture are assessed. The extent of this
'sportification' is a measure of the rationalization of sport through
processes of legitimization which include the increased democratization of sport.

The most fundamental characteristic of the monolithic social structure of elite international or representational sport which results
from sportification is its instrumental rationalization. The totaliza-
tion (Donnelly, 1996) of international sport indicates that for the
principles of performance, profit and prestige, virtually every basic
component or element of sport has been rationalized to the ultimate
degree for reasons of efficiency and effectiveness.

A notable historical trend of the sportification process has been the
increasing democratization of modern sports. For example, there are
very few Olympic events that are closely linked to social class per
se and there is a marked increase of women participants and events
in which they can compete. Although in the past, even well into the
twentieth century, contests between black and white people, such as
boxers in the US, were not permitted and apartheid operated in South
Africa. Those who have been excluded from sport in the past have
organized their own regulating bodies, such as the Special Olympics,
the Paralympics, the Gay Games and various 'Senior' Games, for

example in the US. Sport has enormous democratizing potential and possibilities for providing opportunities for excluded populations. The democratization of sport cohabits in a state of tension with its inequities and exclusions that are all part of the globalization of sport in modern and postmodern times. The globalization of sport is not an entirely new phenomenon, of course, as was evidenced by the diffusion of British sports throughout the world, the development of international sport federations, and the establishment of the modern Olympic Games. Sport was already a worldwide phenomenon by the beginning of the twentieth century (McIntosh, 1971: 95). However, in more recent years, the growth of the media and the media sport nexus and the advent of satellite and Web-based communications have greatly enhanced this totalizing phenomenon.

There are counter arguments to the thesis that sport has been totally globalized. There is evidence of local resistance; folk sports and forms of recreational sport survive in the face of powerful global economic and cultural processes. Embodied sporting practices are dynamic and always subject to change and transformation in relation to local and global actions. Sporting activities are constantly being modified as the conditions of play are negotiated through relationships and processes that involve a combination of players, managers, administrators, owners, media personnel and spectators, although the geography of power often presents significant imbalances as in the tension between fans of English Premiership football clubs and the multimillionaires from the Middle East, Russia and the US, who are increasingly buying them. All sports are historically produced and socially constructed and while the most prominent cultural forms of sport embody systems of dominant meanings and practices, new sports and sporting practices are continually being invented which may generate forms of resistance and/or offer alternative structures and subcultures. In this sense, sports constitute contested cultural and social terrains.

One of the fields of resistance is in what are called extreme sports or adventure sports, alternative sports, action sports, panic sports or whiz sports. Many of these sports are typically characterized by risk, speed and danger and a desire by participants to maintain control of their bodies sometimes without the intrusion of formalized administrative structures and hierarchical supervision. Many participants in such sports express a rhetoric and follow norms that are anti-establishment and often transgressional in their nature (Rinehart, 2005) These sports are sometimes practised in public spaces such as the street on areas outside blocks of flats and in underpasses and might be considered as modern folk sports, given their grass-roots origins and local variations. Some participants have resisted commercial cooptation and the interest they have attracted and others maintain parallel forms of non-commercial, participant-controlled activities; there are questions to be asked about the dynamics of cultural production and transformation and about sport as a game occurrence, a playful, ludic activity, an institutionalized game, a social institution and a form of embodied social involvement (Honea, 2004). Resistance to the monolithic march of globalization and commercialism and games on the margins raise questions about what is a sport and what is not.

The literature on the historical transformation of sport points primarily to the shifts from spontaneous play to regulated games and sports towards a much more detailed, controlling set of regulatory frameworks through which sporting bodies are disciplined. The global reach of sport may also offer more opportunities to what remains identifiable as sport in all these diverse body practices, which, however closely regulated they are, may retain elements of play and enjoyment. Ideas about post-sport make much more of the possibilities of the transformative powers of sport.

Whereas traditional sport practices often contain and discipline enfleshed bodies as resources to be deployed towards the attainment of external goals, which are largely defined as competitive and performance-based sport outcomes, and fulfilment of cultural-institutional discourses, post-sport practices depart from this body-as-resource schematic. Post-sports, which tend not to include mainstream sports like baseball, basketball, football, cricket, athletics and tennis, are conceived of as moral, reflexive, community-oriented, green, spiritual and anarchic. Post-sports to which the academy became more attentive in the 1990s include activities like snowboarding, skateboarding, BMX riding and wakeboarding. Such pursuits became infused with a counterculture credibility of making up rules and pushing the body practices to the limits; different limits though from more conventional sports. They are called 'extreme' because they often feature a combination of speed, height, danger and spectacular stunts. Although levels of danger vary widely, there is always an element, described by participants as an 'extreme' factor which, it is claimed, causes an adrenaline rush. Such sports include outdoor sports that might have been traditionally seen as more conservative in their cultural associations, like whitewater sports including rafting, kayaking and canoeing, mountain biking, rock climbing, trail running and camping, all of which have become cool. Post-sport is presented as not only providing excitement and a challenge to the rule-governed constraints of mainstream sport but also offering a more democratic corporeality of sporting practices (Wheaton, 2004; Wheaton and Beal, 2003). Methodologically, studies of extreme sports have sought to provide 'a "dialogue" between practitioners and academics' (Rinehart and Sydnor, 2003: 8), drawing upon the idea of a co-production of knowledge and experience and a more egalitarian relationship between the academy and practitioners, although participant observation has always been a feature of sports research, especially in sociology (Woodward, 2008b).

They do take on and may even pass as mainstream sports forms and embodied sporting practices such as swimming, running, cycling and what Wheaton (2004) describes as the residual elements of modern and modernist sport, but their individual or collective engagement and lived experience bear little similarity. Post-sports are thus conceived of as anti-commercial, cooperative over competitive, rejectionist of advanced material technology, socially inclusionary rather than hierarchical, process-oriented, holistic, and internally differentiated in their orientation and engagement and linked to anti-globalization rather than embracing its excesses. A post-sport physical culture values human spiritual, physical and emotional materialization through athletic embodied sporting practices, beyond medical-technical or power and performance criteria.

Sociological research on the role of sport in society has been based on the idea that the organization, institutionalization, administration and promotion of most amateur, elite and professional sports practices in the twentieth century reflected mainstream cultural views that sport could be used to translate key social values and ideologies to youth. Research has largely demonstrated that these ideas reaffirm established classed, gendered, racialized and religious lifestyles. The panorama of widely practised and institutionalized sports confirmed patriarchal, techno-capitalist, modernist styles of living and the ways of being within a social order. Sociological research on the physical cultural practice of sport since the late 1980s has, however, documented a growing list of sports, activities and leisure pursuits which have counter-hegemonically begun to resist the tendency to define sports in such traditional ways (Rinehart and Sydnor, 2003; Wheaton, 2004).

The overlap between the competitive nature of mainstream sports and their powerful connections to hegemonic masculinity may continue to be problematic for those who promote post-sport as expressed in these extreme sports. Similarly, the start of the twenty-first century sees some attempt by the regulatory bodies of sport, such as the IOC, to incorporate some such sports under its umbrella as Olympic sports. Whether extreme sports will retain their resistance to regulation and their exuberance remains to be seen, but this is another plane in the exploration of sport and of the meanings that are attached to embodied sporting practices which is important and noteworthy for the project of this book and, most significantly for the experience of sport, whether it is post-sport or sport.

Conclusion

The discussion in this chapter has traced some historical transformations in the movement through the transition from ancient to modern sports and postmodern in order to abstract some characteristic features of sport and to explore how sports are and have been defined, in terms of which activities and practices are included and what are the distinguishing features of sport and sporting practices. There is an extensive literature which engages with the problems of definition and with the differences between play, contests, competitions, games and sports, and there are distinct strands of agreement between all commentators. Each notes the importance of physicality and embodied activity that sport involves and the centrality of rules and regulations, whether these regulations are the mark of the move from the spontaneous play to the organized, structured play that constitutes sport in modern times or a necessary component of all play. Post-sport theorists acknowledge rules, if only in the subversion of established and traditional regulatory frameworks, although their arguments about rule breaking may often mean the reconstitution of different sets of rules. Whichever approach is adopted, sport is about bodies in play and some framework of organization through regulatory bodies, both on and off the pitch. The genealogy of sport, although it does not present a linear narrative of change, does offer a distinctive field of enquiry and of experience. There are stated features of sport, which constitute sport as an identifiable field of identification, of experience and of study. One of the more problematic areas of agreement in the search for definitions is that sport involves play and always in some sense 'ludic'. Ludic might cover the pleasures of sport, although it could also be seen as trivializing the deep affinities and identifications and the corporeal pleasures that sport can bring. The idea of the ludic is deconstructed in a variety of ways, but is rarely linked to another distinguishing feature of sport which is underplayed in most accounts, which is the inequality that underpins sport, whether ancient or modern. Some post-sport accounts seek to address sport as a route through which marginalization and social exclusion can be challenged and sport has always offered a means of social advancement for disadvantaged young men in particular, however fantastic some of the dreams of success that have been invested in sport may have been for many such young people. There is some discussion of sport's democratizing possibilities and invariably acknowledgement of its universal qualities (everyone has to agree about the rules and play the same game even if the game is played on the street), but there is less questioning of which embodied selves have the leisure to enjoy play or for whom contest and competition constitute play and even fun.

The underlying differentiations in operation in the classificatory systems are mostly assumed and unstated and constitute aspects of the field of sport which are less explicitly expressed. When it is suggested that a history of women's sport would be welcomed but there is little of it to report, there is no questioning of whose sport as well as which sport women might have been playing. Similarly, whilst there are spatial specificities in the practice of particular sports and in their historical development, the only transitions are temporal, from ancient or folk to modern, rather than initiated by particular peoples in particular places. Post-colonial sport is usually in dialogue with the colonial import rather than indigenous, which is not to say that global sport has not been transformed by those who play it often with far greater skill and success than its colonial instigators ever did.

The narrative in the history of sport tends to be linear in the move from ancient to modern and differences are understated in many of the stories that are told about what makes sport.

What is classified as sport has undergone a vast transformation into what is now called modern sport. It can also be argued that some of the major transformations have taken place much more recently than the transition from ancient to modern sports and that over the last half century there has been a major change in the organization of sport and especially through the commercialization of sport and the restructuring effected through the vastly increased role of sponsorship, technological change. These changes are addressed in the coming chapters, starting with some of the differences and exclusions identified above by exploring the status of the embodied identities of some of the bodies on the margins in sport. Processes of inclusion and exclusion bring together the bodies that regulate sport and the bodies that participate.

4
Equalities and Inequalities: Diversity and Neo-liberal Dilemmas

Introduction

This chapter looks at equalities and inequalities and at the policy field of diversity and social cohesion as particular aspects of the links between regulatory bodies and the bodies that are regulated. Sport has proved attractive to neo-liberal governments because it offers opportunities for promoting social inclusion, with the promise of a level playing field where all are equal and all that matters is playing the game by universally accepted rules. Sport might offer hope for social transformation and bring in those from the margins. As has already been suggested, a focus upon bodies and their specificities can be troubling for the embodied selves situated on the margins, who are the targets of policies of social inclusion and diversity; there is always the danger of being reduced to the body and to body practices at the expense of intellectual achievements and possibilities. Allied to this is the political conservatism that has been associated with sport. Sport is an aspect of cultural and social life that has strong links to conformist traditional values that might be seen to resist change. However, sport is a target of government interventions that have been targeted at social inclusion. This chapter addresses the strategies and practices of Foucault's notion of governmentality, for example as applied by Nikolas Rose (1999), which includes those that have been adopted to promote greater social inclusion and cohesion through sport. In sport, regulatory practices have involved and targeted different social, ethnic and cultural groups at different historical moments, where the

healthy person, and the 'good citizen', is constructed as the physically healthy individual. This is explicitly stated in policy documents
such as the UK government's *Game Plan*:

> Sport defines us as a nation. It teaches us about life. We learn self
> discipline and teamwork from it. We learn how to win with grace
> and lose with dignity. It gets us fit. It keeps us healthy. It forms
> a central part of the cultural and recreational part of our lives.
> Sport and physical activity can help the Government achieve key
> objectives. Crucially, it can help us tackle serious health issues.
>
> (DCMS/SU, 2002: 6)

One way of consolidating the body politic is to promote healthy
bodies of the citizenry. This intervention could work both ways; by
providing healthy, contented individuals and law abiding citizens
with positive investment in the wider community of the nation or
it might involve the reinstatement of existing inequalities and not
impose conformity and compliance. The state might expect that
good citizenship and strong affiliation to the nation are thus rein-
stituted through sporting and body practices and identifications.
The tradition is one of the healthy citizen of the nation state, but
the status of the nation has been transformed, none more so than
in sport which offers a multiplicity of identifications for its followers, with 'local' teams having no local players nor even players
of the nation where the 'local' team plays, as is particularly often
the case in English Premiership football, which also attracts more
fans outside the UK, across the globe, for example in Africa and
China, than in it. In changing times and with the reconfigurations
of place wrought by globalization, sport offers a means of providing identifications that cross national boundaries. Globalization and
de-territorialization have both negative and positive possibilities, but
sport still permits strong identifications with the nation and a sense
of belonging which are attractive to the agencies of governance,
and creating rational, self-regulating citizens who have investment
in their communities is especially attractive as a development of
neo-liberalism in its social democratic manifestations.

What sorts of subjects are targeted by the practices of neo-liberal
governance in sport? How might people see themselves in relation to
these practices? As was argued in Chapter 3, sport is not always a level

playing field at all and is characterized by significant social divisions and inequalities that mirror those in the wider society as well as reconstituting some that are more particular to sport. Participation in sport may reinforce inequalities as well as widen participation. Sport has been seen as an appropriate strategy for self-improvement by black and working-class young men, for example, in baseball, basketball, athletics, boxing and cricket, with its colonial legacy and class-bounded traditions of 'gentlemen' and 'players'. Such inequalities can be reinforced by the channelling of black, working class and marginalized ethnic groups into sport or, on the other hand, they might provide opportunities for the exercise of some agency through resistance (Carrington, 1998). Those who promote theories of post-sport emphasize both its liberatory potential for resistance to exploitation and the maintenance of inequalities and its associations with those on the margins (Rinehart and Sydnor, 2003).

Sport, because of its associations with play and pleasure, offers a useful site for the promotion of self-governing citizens. People enjoy sport and sport invokes strong affiliations and a sense of belonging which makes it an attractive site for the promotion of citizenship and, to all governments, for the promotion of national identifications. John Hargreaves suggests that sport's strong appeal 'resides in its ability to harness and channel personal needs and desires – for health, longevity, sexual fulfilment, sociability...in a vivid, dramatic, aesthetically-pleasing and emotionally-gratifying expression' (Hargreaves, 1987: 158). The claim that sport is implicated in the fulfilment of sexual desires requires some deconstruction and might, as a universal, be overstating sport's attractions, but this quotation does condense some of the most powerful appeal of sport. The affective appeal of sport and its deeply felt and widely experienced pleasures can be underestimated by an overemphasis on its rule-governed disciplines.

Sport is also global and presents multiple points of identification which transcend boundaries of the nation state. This does present some dilemmas for the integration of liberal citizens of the state, however, as illustrated in the infamous 'Tebbit test' of national allegiance, when the British Conservative Minister Norman Tebbit, reflecting upon national identity and questioning the allegiance of immigrants in the UK in 1990, suggested that a crucial aspect of national citizenship in a multi-ethnic society like the UK was supporting the national team. In the example of his 'cricket test' this meant England and not, for example, the current visiting team from Pakistan, a country with which a significant number of UK citizens would have strong links, including familial and kinship ties. Although a somewhat crude expression of what constitutes national allegiance, which has more limited resonance in the twenty-first century (Howe, 2006), the 'cricket test' still underpins expressions of racism and social exclusion which remain troubling in the exploration of diversity policies in sport and in the freedom of association and affiliation of individuals which are so closely linked to neo-liberalism.

This chapter starts with an exploration of liberalism and neo-liberal forms of governance in which diversity politics impacts upon sport and the sporting embodied selves who are recruited through the policies of social inclusion. Diversity is located within debates about equality and equal opportunities which are integral to the projects of neo-liberal governance. This is, however, a complex nexus and it can also be argued that neo-liberalism with its emphasis on individual self-regulation is, in part at least, the cause of the problem (Giroux, 2005; Grossberg, 2005) rather than the solution to the inequalities of social exclusion.

Neo-liberalism: Contexts and dilemmas

Liberalism as a political ideology is founded on ideals of critical, rational thought, self-reflection and the value of self-determination and individual liberty; it is universalist, although liberalism too has a more troubled history through its failure to embrace everyone; all women, working-class men and those in particular ethnic or racialized groups for example, being only lately included with the rights of citizenship and still not in all places and under all circumstances. Liberalism evokes a unitary political subject who has a right to freedom and shares this right with other subjects who are alike in this respect. The role of the state is to ensure access to liberty and self-determination and to provide a framework in which people can make free choices. Liberalism has at its heart a number of interrelated tensions, notably those between equality and freedom, between equality and hierarchy, and between sameness and difference. First, if people are free to do as they wish, this may cause inequality and inhibit autonomy and the ability to make free choices; one person's freedom may be another person's exploitation in spite of attempts to differentiate between the two. Second, equality of opportunity is almost inevitably in tension with equality of outcomes. Liberals may claim that liberalism only ever mandates equality of opportunity, but persistent inequality of outcomes inevitably suggests that opportunities are not actually equal. There is a limit to what can be explained by claiming that inequalities arise from unequal natural abilities, although it has often been 'nature' that has been invoked to explain unequal outcomes. In fact, pursuing the opportunities open to one as an individual can lead to a structural hierarchy of unequal outcomes which itself constrains equality of opportunity at a collective level.

Lastly, the ideal of equal rights and obligations as a citizen, in relation to the state and other citizens, implies sharing the same values about what rights and obligations are, that is sharing a moral worldview. Different worldviews can generate different conceptions of what is rightful and obligatory for citizens. This is another instance of the underlying tension between sameness and uniqueness. If liberalism demands, if liberalism can be said to make demands at all, that each individual has the right to autonomy and self-determination, then everyone is the same in this respect.

However, at the same time, ideas of personal autonomy lead quickly on to ideas of personal identity and the desire that one's selfhood be recognized by others, which implies differentiation and distinction (Taylor, 1992: 43). In some ways, this might be called a Romantic gloss on Enlightenment liberalism, which also suggests a concern with the special character or spirit of persons, nations, places. If identity is relational, it is marked by what distinguishes one person or one group from others and how each is constitutive of the other. The autonomy of the self can only be realized when others recognize you as distinct and thus different. The autonomy that all individuals desire, and that liberalism vindicates as the key value which is shared equally, is paradoxically a dependent condition that depends on differentiation.

Another linked tension that underpins liberalism is whether liberty is a natural phenomenon in which governance should be minimal and non-interventionist, a view suited to classic liberalism; or whether it is an elusive construction that needs to be carefully and diligently nurtured. If liberal governance means providing protection

against other people's wilful exercise of their 'rights' by an inter-
ventionist state, which as in social democratic regimes cannot only
regulate markets but also guarantee rights, and thus difference, for
those who have been marginalized or excluded such as ethnic minorities and women, it implies an approach more consonant with a civic
republicanism that emphasizes the individual's place in a collective
(Heater, 1999).

To some extent, there has been some exaggeration of the ubiquity
of neo-liberalism. On the one hand repressive regimes remain,
whatever the opening up of markets and, on the other hand, neo-
liberalism takes multiple forms some of which would have been
described as socialist or social democratic: states still intervene in wel-
fare programmes as well as, if not more so, in markets, especially as
demonstrated in the credit crunch of 2008 and subsequent global
economic recession. However, the social aspects of neo-liberalism
might be deemed a luxury that is unsustainable at times of market slumps. There are different strands of liberalism with different
emphases and neo-liberal economics is not the only aspect that has
to be considered. Romand Coles argues that political liberalism was
no longer the dominant ideology in the USA in the second Bush
administration (2005) and an increasingly anti-democratic politics
was replacing the ideals of equality that underpin liberalism through
the right-wing backlash of the early twenty-first century, although
a critical de-construction of neo-liberalism and an exploration of its
component parts, including engaged activism, might offer a more
optimistic reading of its democratic possibilities.

Since the establishment of liberal thought, usually traced back to
the Enlightenment, regimes based on liberalism which have evolved
into what is still an increasingly ubiquitous form of neo-liberalism
have dealt with these tensions in a number of ways. The global
economic recession towards the end of the first decade of the twenty-
first century highlighted particular tensions that were focused upon
the relationship between economic and social values and the primacy of the market within neo-liberal discourses. Such tensions were
expressed in the disjunction between unregulated free markets and
especially banking and greater state control and regulation border-
ing on state control. Tensions have been expressed in different ways.
The most common approach historically has been to confine political rights and the rights and freedoms enshrined in liberal ideals to

particular groups of citizens who have been predominantly white, upper- and later middle-class men. Various sorts of others have been excluded or limited, usually by virtue of a supposed incapacity to exercise such rights or discharge the accompanying obligations. Thus, women, children, the illiterate and diverse ethnic minorities have all, in different places and at different times, been formally or informally excluded from the basic rights of citizenship, such as voting, not to mention other types of liberties. These histories inform the transformations of neo-liberalism, an ideological form of governance that can be seen as increasingly dominant across the globe, which is not to deny the perseverance of repressive regimes and masquerades of liberalism nor its own dilemmas. Neo-liberalism has taken the concept of equal opportunities on board in its defence of the benefits of meritocracies that could emerge, by rationalizing the advantages, not only to the economy, but also to the wider political, cultural and social terrain of social relations and social order.

Economic rationality is fundamental to the neo-liberal project although the economic is extended to encompass all aspects of social, cultural and political life. Wendy Brown argues that neo-liberalism

> entails submitting every action and policy to considerations of profitability . . . human action [to] entrepreneurial action, conducted according to a calculus of utility, benefit, or satisfaction against a micro-economic grid of scarcity, supply and demand and moral value-neutrality . . . Neo-liberalism normatively constructs and interpolates individuals as entrepreneurial actors in every sphere of life. Moral autonomy is measured by their capacity for self-care . . . A fully realised neo-liberal citizenry would be the opposite of public-minded, indeed, it would barely exist as a public. The body politic ceases to be a body but is, rather, a group of individual entrepreneurs and consumers.
>
> (Brown, 2003: 3–5)

Brown's argument draws on utilitarian views of the liberal state and a conceptualization of the state as a collection of individuals and also points towards the possibilities of Foucauldian regulatory practices which provide one route through understanding the operation of policies of diversity and cohesion which are explored in this chapter.

The politics of diversity and cohesion

Diversity has emerged as a significant concept in the formulation of government policies across the Western world. It has a complex, if relatively recent, history, however. Much of the debate has focused upon the politics of race and anti-racism. For example, in the USA this has often been characterized by affirmative action debates following the 2003 Supreme Court overturning of the 1996 Court of Appeals ruling that a race-conscious affirmative action programme at the University of Texas Law School could not be justified on the grounds of the desire to promote diversity. In the cases of Gratz and Grutter, in 2003 it was established that promoting diversity could provide the main justification for affirmative action policies. Thus diversity became a central concern of equality policies. This decision took place in conjunction with moves in the corporate sector towards the growth and enhancement of diversity management (Wrench, 2005). Affirmative action would have particular articulations within sport where the characteristics of physical competence and competition might present difficulties. Diversity was on the agenda and in the language of governance; more than that, it seems to have become ubiquitous, for example, in the USA and in Europe, where it has emerged as a political priority. The concept of equality has been pivotal to the European Union's *Charter of Fundamental Rights* which enshrines a range of equality principles (Shaw, 2005). The Articles of the EU recognize not only cultural and linguistic diversity but also the necessity of combating discrimination on grounds of sex, racial and ethnic origin, disability, age, religion and sexual orientation, so that equality policies combine to ensure formal anti-discrimination, promote significant equality and manage diversity (Bell, 2003).

These policies not only are widespread, but also clearly have implications for the management of diversity and promotion of equality in sport. However, the theoretical underpinnings and rationale for this explosion of diversity require fuller consideration, in order to explore the impact in the field of sport. The promotion of diversity tells different stories, which are implicated in the tensions and ambiguities of neo-liberal thought. On the one hand, what is called diversity might appear to reflect the concerns of those who are on the margins in a view that is supported by the claims of social movements and marginalized cultural groups and located within the terrain of political activism and the promotion of human rights. On the other hand, diversity could be a managerial policy and a modality of the neo-liberal governments that have also become ubiquitous and seek to maximize efficiency and economic and financial success. In practice, the two different versions of the politics of diversity are, of course, closely interlinked; political activists and marginalized groups meet the business sector and managers of human resources in a variety of different spaces. Social inclusion and the promotion of diversity make for good governance and contribute to the versions of managerialism that normalize neo-liberalism as a dominant form of government, although, in practice, the strategies adopted for making diversity fit may be different in the different areas of governance in which it is implemented. Sport is big business and a part of the global economy in many ways that are no different from other enterprises; sport is also subject to the vagaries of the market and to the effects of the credit crunch and global economic downturns, like any other business. Sport is subject to the same criteria of profitability and efficiency that other corporate groups are, but it is also a distinctive field, in no small part because of those embodied subjects who are participants in its narratives and projects. Equality and equal opportunities are played out in different ways in the field of sport.

Making sense of equality and difference

The debates about equality within contemporary neo-liberal political theory which inform policy-making have a considerable impact upon sport. Liberal egalitarian theory is characterized not only by a debate between equality of opportunities and outcomes, but by different sorts of equal opportunities. The notion of equal opportunities is problematic in different ways. What sort of opportunities? How can people be given such opportunities and what happens to those who are unable to do anything with the chances they are given? Meritocracy has attractions within competitive sport, although in terms of widening participation in physical activity it might be counterproductive. The meritocratic implications of equal opportunities might lead to an extremely unequal society with large disparities between a highly paid successful elite and a group of disadvantaged people who come to be classified as victims of their own lack of ability. As John Rawls has argued, this sort of equality might be the 'equal chance to leave the less fortunate behind in a personal quest for influence and social position' (Rawls, 1972: 108). Another dimension of this discourse is that which emphasizes effort and the desire to succeed rather than talent, but this has limited resonance within sport except within charitable discourses of social inclusion (Woodward, 2007b), although effort is clearly important in the promotion of healthy bodies through physical exertion. The debate has shifted within the wider political terrain, but in those places where sport is part of the citizenship agenda, participation emphasizes the role that individuals play in taking responsibility for themselves. The free market is seen as the best route to social egalitarianism and justice, but it is tempered with the recognition that people need extra resources on the grounds of need, as, for example, in the case of disability. Sport differs from other fields where need is defined as encompassing social and cultural disadvantage on a much wider scale, although this does raise questions about what could or should be the outcomes of equal treatment of those who have been so classified as socially excluded or disadvantaged. Equal opportunities in sport may include talent scouts seeking to enhance the pool from which elite athletes might be drawn, a role served by elite universities and clubs in the past. In this case the outcome would be access to training schools of excellence and selection to compete at the highest levels. This raises questions about difference; would women, men and disabled athletes train together, have separate academies or training centres? Would they be equally resourced? These questions arise, however, because of the unequal field of play so characterized by the protection of privilege and the conservatism of sport. They are also embodied issues, situated in a field that is transforming, through the advances of techno-medical interventions and through training regimes that are implicated in social changes. Such 'equal opportunities' operate in tandem with social inclusion programmes which, although apparently promoting equality, act more as charitable projects where the outcome could never be entry into the sphere of elite sport. In sport, the problem of gender and the dilemma of multi-culturalism in multi-ethnic societies present significant examples of such tensions. Each of these areas is concerned with one of the debates which has informed discussion of equality and equality of opportunity in the past, namely the tension between equality and difference, although discussion has been most voluble within the field of gender studies, especially within the framework of feminist theories.

Gender equality and difference

The distinction between equality and difference (Scott, 1997) presents questions that have informed the debates about equality that precede those about diversity but are still very much part of its discourse. This discussion has been framed by what in feminist literature has been called 'Wollstonecraft's dilemma' (Pateman, 1989), which suggests that equality and difference represent conflicting and incompatible aims; if you recognize difference, you cannot have equality. The women's movement has engaged with the tensions between at one point claiming equality and at another requiring the recognition of difference. Questions have been raised about who women might want to be equal to, especially given the hegemony of the white middle-class male who is assumed to be the norm of the rational individual at the heart of liberalism. What is also particular about this debate is that it includes a focus upon embodied difference, for example, between women and men, for example, in employment law, in terms of women's capacity to bear and feed children. Feminists have been nervous about recognizing the possibility that difference matters at some moments and not at others and that it might be possible to construct an agenda in which women's distinctive voices are heard without having to resort to a rationale of cultural differences where women might be universally defined as nurturing and hence subordinate to the male norm of rationality. Given the centrality of classificatory systems based on embodied differences in sport the equality difference debate has some bearing on the gendering of sport and the inequalities between women and men that persist, especially in representations of competitive, global sport.

Feminists have made considerable progress in deconstructing the equality difference binary in most productive ways. As Joan Scott has argued, 'When equality and difference are paired dichotomously, they structure an impossible choice. If one opts for equality, one is forced to accept the notion that difference is antithetical to it. If one opts for difference, one admits that equality is unattainable' (Scott, 1997: 765). An alternative position deconstructs the division, so that neither aspect is privileged. Elizabeth Grosz has suggested an alternative to this dilemma which avoids the negative effects of reversing the equation and privileging either one side or the other, by arguing that a better strategy is 'not to privilege one term at the expense of the other, but to explore the cost of their maintenance' (Grosz, 1994: 32). If difference means putting women at the centre, Grosz's approach to diversity aims to deconstruct the centre and thus subvert the dominant, unequal, existing norm, without assuming an additive model. Thus women are not 'added on' to the existing social structures and norms, which would lead to integration into the malestream, but are part of a transformative process.

> This debate, seeking to go beyond the politics of equality versus difference, has not largely been applied to the field of sport, although it has been very important in feminist theory. In policy making and the processes of governance, these debates can be seen as linked to stages in feminist thinking, from first wave feminism which focused upon rights of suffrage and liberal principles of equal treatment before the law, through second wave feminist demands for positive action and sometimes separatism, to a third wave concern with diversity.
>
> (Rees, 2002)

Thus, in sport those on the margins are not added onto or inserted into the current competitive class-based sporting field, but are included in a reconfiguration of that field and its embodied practices.

Multi-culturalism

Multi-culturalism intersects with gender in its concerns with the equal values that are to be given to difference in pluralistic societies. Multi-culturalism has been seen as a desirable outcome of multi-ethnic societies which are characterized by pluralism and equity, where a range of cultural positions could be accommodated, including religious differences. Multi-culturalism offers the possibility of all cultures being equally valued in a society in which different ethnic groups are largely assimilated, but each retains their cultural distinctiveness; difference and equality might cohabit amicably and productively. Multi-culturalism carries much of the conceptual baggage of the equality difference debate, since it presents the dilemma of recognizing differences between groups of people whilst also claiming that each group is equally valued and has equal rights; the discourse of rights also plays a powerful part in these debates. Multi-cultural societies are also, of course, marked by gender and other aspects of differentiation which cut across and intersect with those of ethnicity and culture.

The interconnections between theorizing equality and difference in relation to gender and multi-culturalism within the framework of global neo-liberal governance are illustrated in the field of sport in relation to some of the dilemmas that emerge from promoting sports participation among women whose cultural traditions do not include taking part in sport. Practitioners have tended to stress the benefits of incorporating women into the global mainstream by encouraging involvement in sport, for example in developing countries. This could be construed as a possibly insensitive or patronizing approach that downplays the voices of women in the developing world. As Chandra Talpade Mohanty has argued, a feminist concern with multi-culturalism has to engage with the heterogeneity of peoples and to deconstruct sources of power which govern who is allowed to speak. 'Black and Third World feminists have provided powerful critiques for the way in which Western feminism has defined the universal in terms of the West' (Mohanty et al., 1991: 56); women in developing countries are seen as 'ignorant, poor, uneducated, tradition bound, domestic, family orientated, victimized – in contrast to implicit assumptions about western women as educated, as modern in having control over their sexualities and the freedom to make their own decisions' (ibid.: 57). This is not only about the construction of a gendered Orientalism, where, as Edward Said argued, through discursive, representational systems, 'others' are constructed as alien and outside, so that the Orient becomes the binary opposite of the 'Occident' which is the West that is taken as the norm (1978). The Orient thus becomes both stigmatized and exoticized. It is also an illustration of attitudes within Western countries. Although most feminists are only too aware of such bias in some earlier versions of second-wave feminism, this statement does serve to demonstrate the need to ask questions about whose interests are maintained by equality arguments and to explore the ways in which equality and difference are interrelated and cannot be sustained as binary opposites. Cultural diversity is always implicated in arguments about the principles of equality and the possibilities of greater democratization.

The debate about equality and the promotion of egalitarianism through multi-culturalism has taken a different turn in recent years. The global spread of terrorist activities such as the killings in Mumbai in 2008 have raised the alarm about threats to democracy. Since the terrorist bombings of 9/11 in the USA, Madrid in 2004 and 7/7 in London in the UK, there has been a flurry of obituaries of multi-culturalism and even of neo-liberal politics (Coles, 2005). The optimistic possibilities of a radical review of the concept by Bhikhu Parekh in his report on multi-culturalism (2000) have been confronted with a pessimism that claims that multi-culturalism is practically impossible and theoretically defunct (Gilroy, 2005). Some critical liberals point to the limitations of multi-culturalism because in pluralizing public culture it permits reactionary cultural minorities, albeit under the banner of diversity, to discriminate against women and children with impunity. One such assertive critique is that of Brian Barry who seeks to distance himself from the communitarian aspects of multi-culturalism (Barry, 2001), which, he argues, actually undermine anti-discrimination and suggest that Parekh's optimism serves only to reinstate existing norms.

Pathik Pathak makes the distinction between multi-cultural sectarianism and factionalism and argues for social practice to move beyond difference in the achievement of equality of outcomes through the discourse of human rights (Pathak, 2008) which also has application across the field of global sport. The concept of multi-culturalism still has both resonance and potential as part of the conceptual framework for understanding equality and difference and the governance of diversity (Modood, 2007). Multi-culturalism suggests that social relations and especially belonging are complex and have to accommodate difference and remains a site for activism in spite of its problems.

Sport is a site where such struggles which bridge the local and the global are taking place and where the global exists in the local; multi-culture is not 'out there', it is everywhere. The political discourse of equality and human rights is also part of what takes place in sport. In 2003 the North American Society for the Sociology of Sport (NASS) Annual Conference subject was 'Sport and Human Rights'. As Michael Messner notes, sport has been instrumental in alerting us to the need to 'bring about equity and fairness and to confront core issues of violence, misogyny and homophobia in sports' (Messner, 2002: xii). There is a focus upon discrimination within the nation state but sport offers global possibilities and sport is the site, for political struggle takes place in sport too, for example in Iran where women are seeking to enjoy football and resist the discrimination which has prevented them for so doing (Fozooni, 2008). Women's participation in the sport is seen as so challenging to the political order that a women's team was disbanded early in 2009 because women and men had played together (Canadian Press, 2009). Women and men are not permitted physical contact and Iranian women are not even allowed to watch men's games so a mixed game was clearly perceived as deeply threatening to the gendered social order and a step too far.

The equality difference debate and its translation into the language of diversity inform the strategies and interventions of neo-liberal governance, which play an increasingly important role in the practices of governmentality. The focus of this book is on sport and sporting practices and sport forms part of the wider terrain of governmentality, which is worth exploring further, especially as a means of interrogating the interconnections between regulatory bodies and embodied practices. The promotion of healthy bodies through participation in sport has a place in what Nikolas Rose calls 're-figuring the territory of government' (1996b: 327) where the subjects of governance are divided into those who are 'affiliated' or included and those who are marginalized (Rose, 1996b: 340). My concerns in this chapter are particularly with how this plays out in the policies and practices which target the marginalized. Such policies are often constituted within the transforming and increasingly mobile and fluid arena of 'community' in which governments are confronted by transnational associations and rival nationalisms, although in the mechanisms of governmentality, community is a mobile and contentious category.

Governments, governmentality and sport

Foucault's notion of governmentality which is linked to his understanding of bio-politics has been very productive in the exploration of the interconnections between regulating bodies and regulatory bodies in sport. Foucault's work has been particularly useful because of the way in which he highlighted the body as a key target for the operation of the mechanisms of power, for example, through the technologies of bodies (Foucault, 1977) which he combines to demonstrate how bodies, as 'population', become the target of the diverse apparatuses of the state through the processes of bio-politics, whereby individual bodies are regulated (Foucault, 1988a, 1988b). The concept of governmentality has been developed by later writers (Brown, 2003; Burchell et al., 1991a; Rose, 1996b, 1999). As Rose suggests, governmentality encompasses all those 'fields upon which one might locate investigations of power/knowledge' (1999: 22) and constitutes disciplines through which the conduct of populations can be regulated. Discipline is linked to new ways of thinking about political rule and about agency. As Rose argues, governments 'seek, with some notable exceptions, to act upon these domains by reshaping the conduct of those who inhabit them without interdicting their formal freedom to conduct their lives as they see fit' (Rose, 1999: 23). This has particular resonance within sport because of the centrality of embodied practices and the par-ticularities afforded by sport as a site at which self-regulating cit-izens might be constituted. Bio-politics works in particular ways in sport.

The body is targeted as the site of intervention, in a sporting version of the bio-politics which permits the state to invade the private space of the domestic and the corporeal in order to secure overall benefit (Donzelot, 1980), later construed in a more disembod-ied dimension of governmentality (Rose, 1999). Foucault's notion of bio-power challenges traditional views of the operation of power. As Foucault has argued, 'this bio-power was without question an indis-pensible element in the development of capitalism; the latter would not have been possible with the controlled insertion of bodies into the machinery of production and the phenomenon of population to economic processes' (1981: 140–141). The regulation of sexuality is clearly central to the operation of bio-power and the control of pop-ulations, which has application to a range of aspects of the embodied self, including race, with its particular histories of scientific racism that has often been applied to sporting achievements, gender more widely and disability. The technology of power is concerned with reproducing healthy bodies and social inclusion is promoted through diversity policies.

Diversity policies and practices are themselves diverse. They range from legislation with, in some instances, the threat of coercion or other sanctions for contravention, to economic incentives for compliance or even for creative responses to such incentives. All involve new language, new sets of practices and expectations and whole new categories of employment and job description. The range of mechanisms involved in their transformative potential lend themselves well to a Foucauldian approach focussing on governmentality as involving diverse assemblages of persons.

(Rose, 1996b)

Tensions within sports diversity policy initiatives reflect those in other areas of the multi-faceted complexities of the relationship between the state and diversity, although there are particularities within sport because of its specific alliances between competition, celebrity and community. Diversity poses problems in sport that have resonance in the wider arena, because of the inherent conflict, or at least ambiguity, between the positive aspect of diversity, based on the freedoms that the liberal state cannot deny its citizens, and the negative aspects which are premised on normative liberal notions of equality; between policies that demand both multiculturalism and the celebration of difference on the one hand and anti-discrimination practices that seek to gloss over differences on the other. Diversity in sport, as elsewhere, is concerned with addressing inequalities and marking out differences, making them explicit (and even measuring and recording them precisely) and yet at the same time claiming such differences *do* not matter because they *should* not matter. As Christian Joppke observes, such policies, which he classifies as liberal-democratic, call for 'different, even diametrically opposed responses to both appearances of diversity: abolish it by means of "anti-discrimination" policy or protect and promote it by means of "multi-culturalism" policy' (2004: 451). In effect, they demand 'the simultaneous rendering invisible and visible of ethnic diversity' (ibid.: 451). There is both the widespread celebration of difference and heterogeneity and the homogeneity of a common humanity that shares equal rights and duties that, as John K. Noyes argues of post-colonialism, 'requires a dialectic that can account for the unifying and homogenizing moments' (2002: 274). Such a moment happened in 2009 with the inauguration of the first black president of the USA. The optimism, with which this momentous occasion was greeted by many people across the world, although especially in the USA, is reflected in much smaller moments in sport, when it is difference that makes equality possible. This dialectic has to be situated and to be located within the temporal and spatial specificities of the institutions and cultures within which encounters take place.

Sport carries particular inflections of this dialectic. There is more stress on social inclusion than upon any celebration of difference, but the policies and practices that are increasingly part of the mechanisms through which diversity is promoted also demonstrate the different elements of multi-culturalism upon which Stuart Hall draws in his discussion of the tension between the more radical critical, revolutionary approach and corporate multi-culturalism (Hall, 2001). Whilst not wanting to make excessive claims for the radical potential of sport, much of which is clearly big business and highly competitive, through its very mass appeal and popularity, those involved in promoting diversity could, can and do deliver a radical message.

Sport and diversity

Sport has a history of providing a medium for the regulation of those on the margins, the dispossessed, notably young men, working-class men or those in disadvantaged ethnicized or racialized groups. Boxing remains a route into a livelihood for many migrant young men; early twentieth-century football provided entertainment and powerful identifications for working-class men to keep them occupied on Saturday afternoons as well as continuing to present dreams of personal success for street children in Brazil and hope of being recruited for the English Premiership for many young men in Africa. This is not to suggest that this is a one-way process; it is not a uni-directional system of control, nor has the strategy always been successful for the processes of governance. Those on the margins have also been excluded from participating, banned from golf, tennis and polo clubs or only permitted to play in a separate category, as the 'players' and not the 'gentlemen' of cricket. Bouts between black and white boxers were officially forbidden until well into the twentieth century in the USA. Earlier fights of the black boxer, Jack Johnson, between 1908 and 1915 were understood as challenges to white supremacy with his victory over Jim Jeffries leading to riots. Women have a particularly long history of formal exclusion from most sports and continue to be excluded from some events, such as boxing in the Olympics (which is still the case at the time of writing). Many private clubs still exclude women from full participation and marginalize women on a number of ways, for example in sports like golf.

Changes have taken place, at least in part, because of the commercialization of sport; liberal economics and market forces have promoted more egalitarian policies and practices, especially for men, although changes are taking place in women's sport, for example, with the growing popularity among young women of sports like football across the globe, especially in the USA and in parts of Europe. The US baseball offers one example of change, although more in relation to race and ethnicity than gender, from the twentieth century. Professional baseball major league baseball teams banned African American players in the late nineteenth century, but black baseball players organized their own highly successful teams and leagues (Petersen, 1992). Unsurprisingly, major league professional baseball reconsidered its position after the Second World War and team owners sought to recruit black athletes in order to capitalize on their talent and to attract the growing numbers of supporters of the black teams (Blalock, 1962). With the expansion of professional men's leagues in baseball, basketball and football in the late 1950s and early 1960s the demand for high calibre players greatly increased. The extent of the egalitarian impact can be overstated, of course, since, as Blalock argues, black athletes were often compelled to work for lower wages than white athletes, for example by being given non-bonus contracts (Blalock, 1962). The strategy for including black male athletes did lead to significant returns for some of them and to an exceptionally high representation of black players. Most sport franchises had, however, been unenthusiastic about signing up minority ethnic players and opted for tokenism in the early days. For example, Gerald Early describes the quota system for the signing of black players and the decision-making process for inclusion that was influenced by knowledge of those black athletes who had had experience, for example through college teams, like those at the University of California, Los Angeles (UCLA), which opened up first to black athletes (Early, 1998). Branch Rickey, the general manager of the Brooklyn Dodgers who hired Jackie Robinson in 1947 as the first black player in major baseball league, did so because he knew that Robinson was familiar with the world of white athletes with whom he had competed at UCLA. Not only was Robinson a super athlete, but also he was an acceptable one to the white establishment. The story is complex, however, and even if the impetus came partly from liberal market forces, black athletes have made inroads into the wider aspects of sport, extending to team ownership in professional basketball, even if it has taken time. However, in 1998 only three head coaches in the NFL were African American, although basketball has fared somewhat better with seven African American head coaches in 1998 even though the league was and is 80 per cent black players (Sage, 1998).

The bio-politics of sport: Race and gender

Boxing is a sport that not only highlights the corporeality of sport in dramatic ways, but also presents powerful expressions of the links between the regulatory framework of governmentality and the representations of its embodied practices. Boxing films play a significant role in the configuration of embodied selves in the genealogy of the sport and especially in the constitution of its heroes. Films too are subject to regulatory practices and boxing offers some illustration of the processes through which the two intersect; the boxing body and the regulation of its representation. Lee Grieveson explores 'Fighting Films' to show how racism becomes a 'concrete effect of the play of modern technologies on the life of individual bodies ... the level of population and the way they reproduce' (Grieveson, 1998: 52), using the US response to the black heavyweight champion boxer Jack Johnson (1878–1946). Johnson's dominance between 1908 and 1915 presented explicit challenges to white supremacy, based on the suppression and regulation of that which destabilized the dualisms of white domination. Johnson's biggest fight, 'The Fight of the Century', against James J. Jeffries, a white boxer who had come out of retirement to prove that he could defeat a black fighter, even took place on 4 July, Independence Day, 1910. This period was marked by racial tensions and race riots, with even the most liberal of commentators, like Jack London, demanding victory for the 'great white hopes' who fought against Johnson. Grieveson argues that racial exclusion and sexual politics shaped the institutions of both cinema and boxing, by focusing on regulatory spaces and practices, 'including measures directed at the economic sphere – particularly the control of the flow of commerce and large corporations – and other measures directed at, for example, the control of sexual morality, of the "traffic" in "white slaves", of structures of white supremacy, of the dissemination of "immoral" representations and so on' (1998: 42). Grieveson points to the ways in which cinematic representations of Johnson's body, his sexuality and especially the perceived, created threat of black hetero-sexual masculinity and his victories over white contenders repeatedly challenged racialized divisions in the USA.

Films of Johnson's fights raised demands for censorship on grounds of immorality. However, the governance of cinema depended on establishing a film's social function and its definition of commerce. The Sims Act of 1912 which restricted the interstate movement of films had been passed in response to Johnson's defeat of Jim Flynn, 'the Great White Hope', in 1912, but the act did not suppress the showing of fight films, it only prohibited their transportation across state boundaries. In Grieveson's opinion the Sims Act was a 'literalization of a broader edgy, constant patrolling of whiteness, that, in its intensity, may, speak also to... the necessary centrality and suppression of *blackness* in the making of American whiteness' (Grieveson, 1998: 45). Thus the policing, monitoring and surveillance of Johnson's body and its cinematic representations combine to provide a genealogy of the regulatory mechanisms of racialization and racism, sexuality and national identifications.

UK diversity politics in sport

Policies of diversity and social inclusion in sport have taken different forms. One strand emphasizes containment; fans, especially young men could be a threat to the social order if disaffected. Another more liberal strand stresses the need to develop techniques of the self and to bring in self-regulating 'good citizens' through sport. This merges with the apparatuses of governance that seek to include those from under-represented groups, either for altruistic reasons or more politically on the basis of rights. Such developments have a significant presence in the UK, which has taken the lead in implementing anti-racist policies at football clubs, although this is increasingly apparent in activities across Europe (FARE, 2008). Football attracts a significant fan-and-player case at all levels worldwide. Globally, the FIFA Big Count in 2006 revealed that there were 265 million male and female players and a further 5 million referees, coaches and other officials, a grand total of 270 million people, or 4 per cent of the world's population, are involved in football (FIFA, 2006). In the UK it is a very popular participation sport among women as well as men (Women's Football, 2009) and the UK was at the forefront of using sport, especially football, as the site for anti-discrimination policies and for the promotion of social cohesion (Woodward, 2007b). Women's and girls' football has grown in recent years with more players competing in affiliated competition than any other team sport in the UK, for example. There has been an increase in the number of players, clubs, leagues and competitions since 1993; the number of affiliated players has increased from 10,400 to over 150,000 in 2009. For example, Sport England's Active People survey in 2008 highlighted that 260,000 women and 1.1 million girls play some form of football in England. There are 26 million females playing across the world, of whom 4.1 million play affiliated football, which constitutes a 54 per cent growth since 2000 (FA Women, 2009; FIFA Big Count, 2006). Women may be keen to play football and it has taken over from more traditional girls' sports in schools, but the women's sport does not have the public media presence of the men's game. Nonetheless, football is an enormously popular game among fans and participants.

In the late nineteenth century, with the arrival of mass, male sports participation and the enthusiasm with which the sport was greeted by fans in the 1880s and into the twentieth century, football was promoted as an activity to divert the energies of working-class men from political engagement in which they might challenge economic inequalities and exploitation, leading to the redefinition of sport as commercial entertainment (Giulianotti, 1999, 2005a; Hargreaves, 1986). British football culture, often localized in areas of high urban density and industrialization, developed in the market place as well as through spontaneous community activity. As Neil Taylor argues, most football clubs came out of works teams run by factory owners, 'established in law as private concerns with boards of directors' (Taylor, 2004: 48).

Later in the twentieth century, the governance of football became more specifically organized around direct control through the targeting of fans as 'hooligans' (Williams and Wagg, 1991). The disaster at the Heysel stadium in 1985, when 39 Juventus fans died, was a watershed in the transformation process, as expressed by the Football

Supporters' Association (which later became the Football Support-ers' Federation) in challenging unfavourable views about football fans and providing an argument against the government's unpop-ular football policing policies. Media coverage of the particularly conflictual decades of the 1970s and 1980s demonstrated the con-struction of football fans as deviant (Giulianotti and Williams, 1994), a view supported by the responses of official enquiries into a series of football stadium tragedies, for example in the Popplewell Report (1986) on the Bradford City fire at Valley Parade and, more particu-larly, in the Taylor enquiry (1990) into the Hillsborough disaster in Sheffield, which gave the main cause of the disaster as 'police failure'. In both cases fans had been perceived as potentially dangerous and deviant. Spectators had been prevented from gaining access to the pitch through structures which were organized to safeguard players from pitch invasion in the case of Hillsborough and to prevent fans entering the ground without paying after the match had started at Valley Parade. Although ensuing reports and recommendations have focused upon structural safety at sports stadia, there is still some resonance of the construction of spectators and fans as potentially dangerous and as the cause rather than the victims of tragedy. The positioning of fans as deviant and dangerous has had considerable purchase within the discourse of sport through identification with particular versions of working-class masculinity.

More recently, sport has become the target of policies and interven-tions categorized within the framework of cultural diversity, social inclusion and social cohesion. The promotion of diversity in sport is driven by government policies organized around social inclusion, although the language has shifted slightly from the framework of the inclusion/exclusion binary to a focus on cohesion, especially fol-lowing the Parekh Report (2000), and on diversity. The Department of Culture, Media and Sport (DCMS) cites the 1996 EU Lisbon Sum-mit as the source of its social inclusion policy, of which diversity is one aspect. Member states of the EU have been committed to pro-moting social inclusion and cohesion as strategic goals. As Stephen Wagg argues, the concept of social exclusion has 'a special place in the lexicon of the contemporary Labour Party' (2004: 14) as part of labour's values, promoted in the context of a free market with all that that entails for liberal democracy, defined primarily in relation to the labour market. This has implications for other fields, including sport.

The Social Exclusion Unit (SEU) set up by the Labour government in 1997 published *Bringing Britain Together* (1998) and created 18 'Policy Action Teams' (PATs) which were central to the implementation of this policy. The PAT 10 Report, covering the arts and sport based on the 1999 Collins Report (Collins et al., 1999), which focused on the relationship between sport and social inclusion and exclusion, led to the publication of *A Sporting Future for All* (DCMS, 2000) which was followed by *The Government's Plan for Sport* (2001) with annual updates in 2002 and 2003.

Diversity policies are part of a proliferation of social, cultural and legislative interventions which could be included in what Nikolas Rose calls the 'practical rationalities' (1996a: 173) which permeate everyday life and the myriad ways in which people take on identity positions whereby selves are transformed (or not). It is the ubiquity and extent of these practices and their productivity and potential for transformation which are most useful for an exploration of the particular interventions associated with promoting diversity. Rose defines practical rationalities as 'regimes of thought, through which persons can accord significance to aspects of themselves and their experience and regimes of practice, through which humans can "ethicalize" and "agent-ize" themselves in particular ways...through their associations with various devices, techniques, persons and objects' (1996: 173).

These practical rationalities for promoting diversity suggest benefits to the wider community of promoting the social inclusion of not only disadvantaged but possibly disaffected groups. Home Office concerns with 'strengthening society' through 'community cohesion' (Home Office, 2005) combine with cultural practices in a discourse of equal opportunities which aims more to produce than to police citizens. Policy statements claim that 'Arts and sport, cultural and recreational activity, can contribute to neighbourhood renewal and make a real difference to health, crime, employment and education in deprived communities' (PAT 10, 1999: 5). This is because they 'appeal directly to individuals' interests and develop their potential and self-confidence' and they 'relate to community identity and encourage collective effort' (ibid.: 8). These policies, like those expressed in the Home Office strategy document, *Improving Opportunity, Strengthening Society*, place diversity within a framework which brings together 'community organizations, business and front line public services' (2005: 5). Football clubs draw on particular communities, although increasingly the spatial location and the fan base of major clubs have become disassociated. Increasingly, it is the physical locality in proximity to the ground that is inhabited by those who are not part of the club's traditional, mainly white fan base who are the target of policies designed to combat social exclusion and promote diversity (Bradbury, 2001). Those classified as socially excluded have become a broad category embracing black and ethnic minorities, women, lesbian, gay and bisexual people, people with disabilities, and disaffected young people, a group which must constitute a significant majority if all are included.

Policies which have become embedded within neo-liberal governance are appropriated within three different discourses, of utilitarianism, charity and human rights, which illustrate the tensions of equality and difference and some of the dilemmas of neo-liberalism. The first and most common policy practice is that of utilitarianism, whereby clubs seek to maximize the pool of talent from which players are drawn by running youth academies under the aegis of social inclusion and increasing their profile as a community-based club with good public relations, for example through wearing anti-racist shirts and supporting diversity initiatives, especially those with a significant media presence (Kick it Out, 2008). Charitable discourses are the least transformative and largely involve a media presence for players at charity events and construct the targets of social inclusion policies as victims who are to be helped by the wealthy clubs. The discourse of charity is reproduced through the classificatory systems that aggregate all under-represented groups as, in some sense, victims and vulnerable, even though in sporting terms most girls and women are well able to participate physically in the sport if the facilities were available, and some members of the target group have little interest in this sport. The women's game is largely relegated to the category of a worthy effort. The last discursive formation is that of human rights which frames interventions within the language and practice of human rights with a coherent political message of diversity and entitlement. This is expressed in the activities of volunteer-based groups like *Kick it Out* and *Football Unites: Racism Divides* or in clubs with an explicit political strategy, increasingly which includes mechanisms to combat homophobia, one of the last taboos in diversity in sport.

Conclusion

Sport and embodied sporting practices are both part of the wider terrain of neo-liberal governance and a specific field with its own characteristics. Sporting bodies generate their own versions of neo-liberal governance that are implicated with other aspects of governmentality through which sporting bodies are reproduced, enacted and experienced. Sport is distinguished by notions of fair play, equity and justice that resonate with the underlying principles of neo-liberalism, but it is also subject to the same market forces as other enterprises and organizations and is itself structured around competition which is measured and regulated by physical performance and corporeality. Thus, sport brings together the regulatory bodies of governance and the bodies of those who participate not only in sport but in the apparatuses through which sport is disciplined. Regulatory practices include the flow of commerce and the maintenance of social order through a range of mechanisms.

Its associations with play might make sport seem a trivial arena for the exploration of debates about equality and difference, but sport attracts powerful identifications and historically has been a major channel through which the dispossessed have entered the mainstream, and have been excluded from it. The politics of sport are part of global and more local political tensions and social divisions and sport is subject to the flows and inequalities which characterize the wider arena in which power relations are played out. The body practices with which sporting participants engage do not undermine its political importance, although the associations of sport with specifically body practices can create troubling constraints for those who use sport as a route out of oppression or social exclusion. The whole notion of sport as a vehicle for the promotion of social inclusion and cohesion is not, of course, without its own problems. An examination of the equality, difference and diversity debate suggests that sport can indicate some of the tensions and contradictions within the wider political field and suggest some alternative ways of thinking about social inclusion as a positive outcome for interventions. What sport has done and continues to do is to make very public some of the contradictions and inequalities that persist in neo-liberal states. Whilst the sport media sponsorship nexus might create its own inequalities it also provides the means of making visible and, by implication, invisible, who is included and who is not. Sport and sporting bodies are very visibly 'put into discourse' through media coverage, which is what has also made sport and sporting success a particularly attractive medium for the promotion of the politics of resistance, as well as integrative policies of cohesion espoused by neo-liberal governance in the making and remaking of good citizens.

The equality and difference conversation permeates multiculturalism and the politics of gender through the idea of fairness and neutrality, but also highlights the neo-liberal dilemma of intervention and how such practices could be framed, for example, as the enabling strategies of agency, fair play and impartiality or within the context of rights and in particular human rights. Sport is characterized by fair play and presents possibilities for the enhancement of rights within a discourse of impartiality and justice.

Diversity has provided a third way and an alternative to the equality difference binary. A focus on rights presents a productive strategy for recovering the political potential of diversity politics as a means of rescuing diversity from the cosmetic masking of inequalities, for example through the discourses of charity or market-based utilitarianism, but the conceptualization of corporeality that underpins this approach remains discursive, with limited attention given to the lived experience of the embodied selves that are the target of such policies. Another approach to the interrelationship between regulatory bodies and the bodies that are regulated is explored in Chapter 5 through feminist phenomenological ideas of bodies as situations and situated bodies.

5
Embodied Selves: Situated Bodies, Bodies as Situations

This chapter highlights the centrality of embodiment in sport and explores some of the processes through which selves are forged through participation in the embodied practices of sport. As was demonstrated in Chapter 4, sport has been marked by the evolution of regulatory practices, for example in the tradition of Juvenal's *mens sana in corpore sano*, which at different historical moments have been applied to different social, ethnic and cultural groups, where the healthy person is constructed as the physically healthy individual. Governments, like that of the UK, in their policy statements equate the healthy nation with the healthy body, which in neo-liberal governance is translated into the self-regulating, healthy citizen. The target group of those under-represented in sport has been expanded to embrace a range of people who have been grouped together in an ambivalent and contradictory grouping of embodied selves on the margins of mainstream sport and society. The language used is often that of diversity, reflecting some of the theoretical shifts in debates within neo-liberalism about equality and difference, although the target group is constructed as homogeneous in its need for healthier bodies. Interventions have been framed in different ways; they have sought to control the disaffected as well as seeking to promote a more inclusive society, both of which approaches have been underpinned by a trust in the benefits of the healthy, fit body. The emphasis so far has been on the production of categories of person through such apparatuses of governance, rather than the people who are so classified. This chapter addresses two problems related to this; first, how to deal with the material lived body that is so targeted and move beyond the idea that the body is a text that is inscribed and, second, how to rethink the possibilities of agency and resistance, especially when the embodied selves who are the focus of the promotion of healthy minds and healthy bodies are those often classified as 'on the margins'. Although structural factors may indeed situate those so categorized as outside the mainstream as disadvantaged or even deviant, how do they situate themselves?

This raises the need to engage with rethinking the body, for example, in relation to the devaluing of the body in relation to the mind as is lived in the experience of social disadvantage which operates in contradictory ways in sport. This need to rethink the mind body binary has been addressed in phenomenological accounts which challenge the mind body dualism (Bourdieu, 1992; Crossley, 2001; Merleau-Ponty, 1962) and applied to sport (Wacquant, 1995a, 1995b, 2004; Young, 2005a) and the centrality of body practices in the making of gendered identities (Connell, 1995, 2002; Young, 2005b). This chapter explores the possibilities of different aspects of embodied selves and looks at how bodies are lived and situated and at bodies as situations (Beauvoir, 1989; Moi, 1999). Individual and collective endeavours in sport focus upon the material body. This chapter presents the argument that bodies are both situated by social, cultural and political factors and are themselves situations, in order to accommodate all elements of the materiality of embodiment, thus combining analyses of embodiment with those of regulation. The chapter starts with some discussion of the politics of sport and some of the ways in which sport can be seen as a site of resistance where agency can be exercised, before examining some theoretical positions that might afford some reconsideration of the tensions between creativity and constraint. Phenomenological accounts that have been applied to sport are expanded to include feminist developments which can accommodate the gender differences that, argue, are insufficiently explained in many existing applications of phenomenological theories. This chapter picks up on the embodied selves on the margins, who are the targets of the diversity policies discussed in Chapter 4, and focuses upon the gendered body as situation.

The politics of sport

Although there may be moments when journalists or even politicians demand that politics be kept out of sport and that the two should never be mixed, this is a largely discredited view in the academy. Sport studies scholars and sport sociologists recognize the interconnections between sport and politics and the idea of sport as a site where power geometries are in play (Tomlinson, 2007). Sport has become a focus for political interventions, at least insofar as it is governments who often seek to promote sport and political parties which might compete to identify themselves with such promotion. Sport also has a long political history and sport and politics have been closely enmeshed at diverse points and in different places over time. The spaces in which sport has been played across the globe constitute an important aspect of the politics of sport. For example, the legacy of colonialism, especially British colonialism, has been particularly influential in some sports in some parts of the world. The dynamic of colonialism and post-colonialism has been powerfully expressed in cricket, for example in the work of Ashis Nandy (1989) on Indian cricket and Mike Marqusee on the decline of England and in some of the everyday resistance and creative energies that go into post-colonial cricket in the contemporary context by Ben Carrington (1998). However, it is C.L.R. James's work, the *View from the Boundary*, that most transformed how cricket and in particular the politics of sport could be understood. James's example is relatively local, namely cricket in the Caribbean, but his message is global. He describes a moment of political conflict which was about national independence but expressed through a game. He combines cricket commentary with political analysis in engaging with the issue of the demand for the appointment of the first black captain of the West Indies cricket team. He thus succeeds in making a very powerful case for the inextricable interconnections between sport and politics. As he says at the close of the Test Match in 1960,

> So there we are, all tangled up together, the old barriers breaking down and the new ones not yet established, a time of transition, always and inescapably turbulent. In the inevitable integration into a national community, one of the most urgent needs, sport and particularly cricket, has played and will pay a great role.

There is no one in the West Indies who will not subscribe to the aphorism: what do they know of cricket who only cricket know?

(James, 1963: 117)

James is highlighting the inextricable interconnections between the embodied practices, here of cricket, that is, what athletes do in the field and the politics of the wider social arena of governance and regulatory practices.

Governments are increasingly recognizing the power of sport as a site for transformation, integration and even inspiration. Sport is clearly political. If, as Nelson Mandela has said, 'sport has the power to change the world' (Mandela, 2007), sport could also be a site of resistance. In the history of sport there have always been big public moments that have framed the politics of sport. The politics of race have long been articulated through sport, especially those sports that have been associated with colonialism or, like boxing, as offering a route out of poverty and into self-respect for black and minority ethnic migrant young men (Ali and Durham, 1975). The black US heavyweight champion Joe Louis's defeat in 1938 of Max Schmeling, who had been used as the representative of Nazi philosophy, has been called a conflict between democracy and totalitarianism (Hauser, 2007). The inclusion of the former USSR in the Olympics in 1952 created a rivalry that paralleled that of the Cold War and was played out in the media. The Black Power salutes of Tommie Smith and John Carlos after the 200 metres in the 1968 Mexico Games attracted global interest and have become one of the pivotal moments in the politics of sport and a key point in African American politics. The deaths at the Munich Olympics in 1972 and subsequent boycotts of sporting competitions by African nations, the USA and the USSR stressed deep political conflicts. Sports boycotts have been used on many occasions, with that of South Africa during its apartheid years being a most significant example. Boycotting was again raised in the run up to the Beijing Olympics in 2008 invoking the politics of human rights. Sport is implicated in the wider political arena in diverse ways and sport cannot escape being a target, as the attacks on the Sri Lankan cricket team in Lahore in March 2009 showed (Lahore, 2009). These attacks in which six Pakistani policemen and a bus driver were killed show how dramatically and seriously sport is imbricated in global politics. The 2009 IPL matches were transferred to South Africa in a move which was heralded as changing the face of cricket (IPL South Africa). Sport and politics cannot be separated.

Politics is not restricted to the arena of governance and the interventions of the state although this is often the site at which politics is most public and seeking redress for injustices through the apparatuses of the state, or in sport the governance of sport is often most relevant. For example, the Olympic Movement's response to the long revolution of social change took different forms: the delineation of amateurism has largely been abandoned altogether; women have been more fully included in the mainstream Olympics, albeit still on unequal terms; racism and racial discrimination are considered unacceptable.

The question of disabled sports presents a more complex problem. Bodies and their classifications are central to these debates as are specificities of embodied sporting practices; what physical acts people can and cannot achieve. Disabled sportspeople are not explicitly banned in any sense from the Olympics, but they have to achieve specific minimum standards of performance. The Paralympics present a compromise solution achieved by the efforts of the disability sports movement. The roots of the Paralympic Games are conventionally traced to the neurosurgeon Sir Ludwig Guttmann, who organized a sports competition for Second World War veterans with spinal cord injuries at Stoke Mandeville Hospital, England, and present a good illustration of political change being wrought through and engagement between embodied sporting practices and then organizing bodies of sport, the IOC, and what then became the International Paralympics Committee, the IPC. Different disability groups and other national organizations worked together to organize the first Paralympics in Rome in 1960 after the Olympic Games of that year. Since the Seoul 1988 Paralympic Games and the Albertville 1992 Winter Paralympic Games, they have also taken place at the same venues as the Olympics. In 2001, an agreement was signed between IOC and IPC so that, from 2012 onwards, the host city of the Olympic Games will also be obliged to host the Paralympics (IOC, 2009). The relationship between the two movements remains contested and the dividing lines between the disabled and the able-bodied, more difficult to determine with embodiment at the centre of the political debate, as I discuss further in Chapter 7.

Politics does not only operate either at the level of the big public sporting displays or through sports' organizing bodies. Mega events have attracted particularly public political expressions. Not only is sport, especially through its spectacular events, the site for public protests, there are also expressions of dissent against the internal mechanisms through which sport is regulated which have been expressed in myriad routine ways, some of which involve less dramatic exposure of sporting injustices, but nonetheless engage with the ways in which the organizing bodies of sport regulate the embodied selves who take part. Resistance also operates in diverse ways, for example through grass-roots resistance as in the anti-racist organizations or through the activities of supporters (Back, Crabbe, and Solomos, 2001, 1998; Woodward, 2007b) or members of local communities (Asians in Football, 2005).

Some of the inequalities that mark sport are explicit in how sport is regulated. There are different rules governing women's and men's sport, many of which have led to the exclusion of women from some competitions. For example, Jennifer Hargreaves argues that women's football was irrevocably damaged by the Football Association's (FA) withdrawal of its support for the women's game in 1921 (Hargreaves, 1994). Women's football was popular in the north of England in the early part of the twentieth century. 'Dick Kerr's Ladies' was the best-known team, and it became the de facto 'England team', playing in the USA and even playing against men at home. They received support from male entrepreneurs, who managed to obtain FA pitches so they played in front of large crowds – 50,000 when they played against Everton on Boxing Day in 1920 (Hargreaves, 1994). Although the FA had provided pitches, by the end of 1921, even though there were about 150 women's football clubs in England and the first Ladies Football Association had been formed, it was not only social conditions in readjusting to the period following the First World War, which affected the women's game, but the explicit hostility of the FA (Williams and Woodhouse, 1991). The FA claimed that 'the game of football is unsuitable for females and should not be encouraged' (Williams and Woodhouse, 1991: 17).

Because football has a historical and cultural past that is clearly linked to male domination, which women's achievements could be expressed as resistance; attempts by women to participate in the sport can be seen as challenging male hegemony. For example, this ranges from the earlier part of the nineteenth century in the UK to more recent engagement across the globe in Muslim countries such as Iran. Babak Fozooni (2008) examines football as a site of social contestation for Iranian women and the recent political controversy regarding women's attendance at football matches within a feminist framework of analysis. She makes a strong case for the political importance of sport as an expression of resistance. Women's bodies and thus women's embodied agency are not only controlled and constrained through the politics of reproduction and sexuality, they are also targeted through the disciplinary mechanisms of sport.

The gendered history of sport illustrates the different aspects of sexual politics since, as Jayne Caudwell argues in her work on sites of resistance in women's football, 'the identity of players has been assumed to be male and heterosexual' (2007: 356). Women footballers have always received comments and innuendos about their sexuality, such as 'a bunch of dykes running round a football; women trying to be men' (quoted in Caudwell, 2007: 358). Caudwell argues that gender identities and sexuality as expressed in sport reflect and reproduce those in the wider society, and uses the example of sexology, a set of theories claiming the masculinizing effects of women's engagement in traditionally male activities, including sport. She quotes a report in the *Daily Herald* from 1935, which describes how 'what was originally, a gentle, feminine girl becomes harsh and bellicose, in all relations to life' (Caudwell, 2007: 358) continue to influence contemporary discourse on women's participation in sport. However, while women in sport still have to combat misogynist and heterosexist comments, sport also offers a site of resistance, whereby

> players who are 'out' and part of teams which adopt the 'in your face' tactic, represent transgression, since they trespass 'man-made' and heterosexual landscapes. Teams such as Hackney provide evidence that lesbians who have entered the 'football field' on their own terms are, to varying degrees, renegotiating and reconstructing this space.
>
> (Cauldwell, 2007: 358)

The physical occupation of sports spaces routinely provides ways of reconceptualizing those spaces and what constitutes sporting practices. In order to re-version those spaces it is necessary to consider how embodied selves occupy them and how, through the politics of and in sport, selves might be transformed. What possibilities are there for reconsidering body practices and embodied selves which permit and explain political action and change? Does the agency implied by political action involve a mind/body split?

Mind and matter: Mind over matter

Bodies are central to sport and its practices and embodiment is pivotal to phenomenology and phenomenological accounts of perception, which engage directly with the problem of the embodied self who embraces the inextricable entanglement of mind and matter. Phenomenology is attractive because of its focus upon lived experience; being in the world is inextricably bound up with the constitution and extension of the body. In phenomenology, the structure of the self is indivisible from its corporeal capacities. Different strands of phenomenology and of Mauss's work on body techniques have been applied to sport, some of which have drawn upon Bourdieu's work, especially on *habitus*.

Merleau-Ponty's work has informed much of the more recent explanation of corporeal schemata that have been developed to challenge the binary logic of mind and body and its associated problems of the question of how to accommodate agency (1962). For Merleau-Ponty, the body has to be understood outside the dualisms that separate nature from culture and mind from body because of the unique status of the body for each person as both an external object and a subjective condition through which it is possible to have relations with the object that is the body. Making the case against a purely objective view of the body, he argues,

> I am not the outcome of the meeting point of numerous causal agencies which determine my bodily or psychological make-up. I cannot conceive of myself as nothing but a bit of the world ... All my knowledge of the world, is gained from my own particular point of view, or from some experience of the world ... Scientific points of view, according to which my existence is a moment of the world's ... take for granted, without explicitly mentioning it, the other point of view, namely that of consciousness, through

which from the outset a world forms itself round me and begins to exist for me.

(Merleau-Ponty, 1962: viii–ix)

Rather than taking this position as an expression of a somewhat naïve understanding of the complexities of consciousness, I think it has particular advantages that benefit a reconceptualization of embodied subjectivity and of the relationship between body and mind that permits both a focus upon lived experience of material bodies and the possibility of resistance and subversion of norms. Perceived as a part of lived experience, the body is a way of being present in the world; the body as a situation encompasses both the subjective and the objective aspects of experience. Merleau-Ponty's theory or perception most effectively challenges the machinic mind body distinction of Cartesian dualism by demonstrating that perception is formed by the interrelationship between the lived body and its environment; acts of perception are what create perceived objects and perceiving subjects. This has benefits in a study of embodied practices in the field of sport because it shifts the emphasis from sport as concerned only with corporeal competences that can be separated from intellectual achievements and from intentionality which can be applied to political action. Merleau-Ponty's corporeal schema also offers an alternative to the notion of an isolated self/body located within, but separate from an external world and presents a means of comprehending how embodied experience works and can thus be seen as embedded in a dynamic relationship with social and cultural practices.

One important aspect of the corporeal schema which is developed is that it is possible to live in the body without constant reflection about the relationship between the subject and the object. Thus it is suggested that we experience our bodies and the external environment in a pre-reflective way that does not require thought at each moment. Routine habits do not require intellectual reflection prior to their enactment but are effected through embodied practice; we learn by doing and do by doing. A corporeal schema is achieved through the acquisition and development of habits, where ways of being in the world become habitual, whether through observation and learning or innovation and creativity. Embodied subjects acquire new habits through dynamic processes of interaction. Habitual acts that do not require intellectual reflection prior to enactment are routine and have particular resonance in sport. A bowler in cricket who decides to bowl a ball as straightforward as a full toss to tempt a batsman does not stop to think before executing the action required to achieve this end. A pitcher in baseball who stops to think is unlikely to be successful. When Serena Williams was serving to win her 10th Grand Slam at the Australian Tennis Open, in 2009, her competence was the outcome of rigorous training but the actual moment of delivery and the actions involved were not the result of reflection on court. Sport is full of such actions and observations by commentators in the event of failure that the athlete is stopping to think or 'hasn't got their eye in' or has 'lost their rhythm'.

Nick Crossley, who has drawn extensively on phenomenological accounts, describes the habitual actions and engagements of lived experience as a situation that he calls a 'moving equilibrium' (Crossley, 2001: 137). Crossley's phenomenological reformulation suggests that agency is realizable through habit; in fact that it is only by becoming habitual that the actions of embodied selves can be seen as agentic. Distinctions are possible between individuals because of their occupation of a particular body in time and space, so that although habits may be shared by many within a particular culture or society, not all experiences of the lived body can be the same. Thus, Crossley maintains the possibility of the individual embodied self and a subject who is capable of exercising agency. However routine the practices, they are experienced by an embodied self who has agency because of the temporal and spatial specificities of the lived body. Crossley's empirical work focuses on the experience of gym culture which has particular application to an exploration of body practices in sport and deploys a phenomenological approach to methodology (Crossley, 2001) which provides a strong engagement with lived experience. He works with Bourdieu's challenge to the mind body distinction and a more fluid understanding of the relationship between agency and social structures, where the first is seen as creative and the second constraining, that Bourdieu sought to develop through his concept of *habitus*. However, as Crossley argues, the habitus as developed by Bourdieu is perilously close to a dualist structuralist account that, implicitly at least, rejects creative capacities (Crossley, 2001: 95).

In order to retain some of the creativity that he sees as inherent in Merleau-Ponty's corporeal schema and to demonstrate some of the dynamics of the relationship between the embodied self and the wider social terrain, Crossley develops a systematic approach to studying the body as part of embodied culture through the idea of reflexive body techniques, which differ from Mauss's body techniques, which have practical and symbolic purposes beyond the body. Crossley's reflexive body techniques are only those techniques directed towards the body itself, its maintenance and modification. Thus his argument yields some insights into how particular societies mark and conceptualize bodies, by breaking down his techniques into core, intermediate and peripheral zones. Crossley finds that all zones, apart from some practices in the peripheral zone, such as the practice of tattooing, are more likely to involve women than men. Body practices are related to the wider social context, for example he locates the tattoo world not only as consisting of individuals seeking particular body projects but also as made up of a commercial culture of tattoo artists, parlours and publications, but offers little critical analysis of gendered embodiment beyond a description of body practices.

Whilst Crossley has advocated a return to Merleau-Ponty's phenomenology to redress some of the problems in Bourdieu's concept of habitus as too constraining and failing to accommodate the notion of agency, Loïc Wacquant offers one of the most well-known and well-developed exploration of body practices in sport which draws directly on Bourdieu's use of habitus to challenge mind body dualisms and provide a powerful account of the making of embodied selves in boxing (1995a, 1995c, 2004).

Wacquant demonstrates from his own experience as an apprentice boxer and temporary immersion in the field that pugilism, and in particular the pugilistic habitus, is achieved through routine body practices of training and sparring. A boxer's moves

> far from being 'natural' and self-evident, the basic punches (left jab, right hook, right cross, straight right hand and uppercut) are difficult to exercise properly and presuppose a thorough 'physical rehabilitation,' a genuine remoulding of one's kinetic coordination, and even a psychic conversion. It is one thing to visualize and to understand them in thought, quite another to realize them and, even more so, to combine them in the heat of action ... Theoretical mastery is of little help ... and it is only

after it has been assimilated by the body in and through endless physical drills repeated *ad nauseam* that it becomes in turn fully intelligible to the intellect. There is indeed a comprehension of the body that goes beyond – and comes prior to – full visual and mental cognizance.

(Wacquant, 2004: 69)

Wacquant does not reflect on the greater likelihood of a renowned Bourdieu scholar and Professor of Sociology at Berkeley, like himself, being more interested in intellectualizing his body practices than most of his fellow pugilists in the gym, but he does make a powerful point about incorporating practices through an almost imperceptible process. Boxing is, of course, a very particular sport, given that the main aim in competition is to render the opponent unconscious and avoid such a fate oneself. However, boxing offers an important site for the development of understanding about how 'we are our bodies' in a form of 'direct embodiment' (Wacquant, 2004: 60) where there is no distinction made between mind and body. The merging of mind and body offers a useful means of understanding, not only how, but also why people subject themselves to such punishing training regimes and participate in what seems to be so brutal a sport. The body practices of boxing go much further than those in other sports in which particular competences integrate mind and body because the intention of the boxer is to inflict physical pain and damage and the risk of incurring such injury oneself. The contradictions and drama of boxing make it a sport that is difficult to understand.

Wacquant's account is persuasive in describing the processes of embodied attachment which take place, for example, in training regimes and the commitment of boxers to their sport. Boxers *are* their bodies and only become boxers through practice and physical engagement. It is not possible to differentiate between mind and body or body and self. Such ethnographic accounts give high priority to body practices and demonstrate how engagement in the pugilistic activity of men's boxing works, especially in terms of the stoicism, resilience and courage of boxers who keep going in spite of injury and pain, which are deeply embedded in traditional masculinity. This is not only what makes a boxer, it is also what makes this version of masculinity. Similarly, as Iris Marion Young argues, women have incorporated ways of doing and ways of being into their comportment, noted in sport through the body practices of 'throwing like a girl' (Young, 2005b). The body practices of men who box are not merely learned; what they do is who they are and these identities are gendered through iterative practice. Yet, Wacquant's account assumes rather than explores associations with masculinity and suggests a universal embodiment which fails to accommodate the specificities of gender. A generic and universal mind/body elision is assumed. What may not be accommodated is disruption. What happens when either mind and body are not one or when everyone is not part of the same system of practical beliefs? This raises a persistent problem with Bourdieu's concept of habitus; it is assumed to work. One of the problems with an account such as Wacquant's is that the researcher is complicit and insufficiently situated in relation to the field of research; disruption becomes imperceptible, if not impossible, in the smooth operation of the habitus.

The immersion of what Wacquant calls the 'observing participant' (2004: 6) may invoke, on the one hand, a perspective that is taken from the subjective position of the researcher and may therefore lack distance or, on the other hand, may provide deeper insights which enrich the findings and create a more accurate picture of the field. Boxing presents a particular site that is strongly configured around gender polarities. The researcher is not only inside the field as a participant, he is also colluding with a particular version of masculinity. Ethnography privileges the insider, placing the male researcher in boxing within a twenty-first-century rearticulation of the *Fancy* (Egan, 1812) privy to disclosures about associated activities, outside the ring and the gym, such as gambling, dog fighting and, in some cases, bare knuckle fighting (Beattie, 1996; Mitchell, 2003). However, these practices and networks are very specifically gendered, in ways that are not always acknowledged.

Yvonne Lafferty and Jim McKay, in their study of the interaction between Australian women and men boxers, draw upon Wacquant's interpretation of *illusio* as 'collective misrecognition' (Wacquant, 2001: 10) and 'collective bad faith' (Wacquant, 1995a: 86). Wacquant suggests that *illusio* is a means of demonstrating that boxers are not deceived by an exploitative system which compels them to sell their bodies to the pugilistic trade but that they are enmeshed in a powerful belief system which holds onto the honour and nobility of boxing. It is hard to see how women could be similarly implicated since the *doxa* of boxing is one that manifests little cultural tradition of women in kinship groups and social networks having boxing 'in their blood', although individual women may give voice to such commitment. Lafferty and McKay cite a statement by an Australian amateur boxer, Mischa Merz, to support their claim that the concept of *illusio* works for women too (2005: 273) and can be applied in the same way as Wacquant does. Another alternative explanation of Merz's statement that boxing is 'in the blood' could be that she is invoking the language of men's boxing in order to be positioned within a discourse which accepts and reinstates the total commitment of the boxer. She is attempting to buy into the language to boxing and to reclaim some of its heroic subject positions in order to be accepted and to 'do' masculinity. There are different cultural meanings attached to women's engagement in the sport which suggest that this could be a gendered *illusio* which is specific to the participants in the context of the wider field in which these meanings are forged.

Some of the difficulties arising from these developments of Bourdieu's theories can be identified as, in the first case, relating to the construction of masculinity which is assumed and could be seen as over-deterministic. Bourdieu's approach fails to include contradiction and ambivalence, especially as manifest in the anxieties that men have to manage. Judith Butler has argued that Bourdieu assumes that the field is a precondition of the *habitus* and that ambivalence is outside the realms of practice (1997), which could certainly be said to be the case for boxing. In the second case, in terms of the construction of femininity in relation to masculinity, the theoretical framework is largely configured in the context of an assumed norm that is masculinity, which is particularly dominant within sport.

Bodies as situations: Situated bodies

Ethnographic accounts such as Wacquant can make significant contributions to an understanding of the constitution of embodied selves from the inside, as a participant. Such an approach demonstrates clearly that sporting bodies are not only symbolic bodies, inscribed by social meanings and visible, visual markers of difference. Material bodies also embody inequalities and experience the limitations as well as the opportunities of corporeality, which is reproduced through body practices. For Wacquant and for Crossley, although partly drawing on different traditions, each focuses on body practices and challenges the mind body dualism so that corporeality is insep-arable from selfhood. Crossley offers a greater accommodation of creativity and the possibility of agency which goes some way towards redressing the imbalance of determinism in Bourdieu's concept of habitus in his phenomenological account. However, his analysis of gender remains descriptive rather than critical which is not neces-sarily a feature of the phenomenological approach and methodology which allows respondents to speak. Iris Marion Young adopts a fem-inist phenomenology which situates gendered experience explicitly and does focus on difference.

The meanings of bodies are not written on the surface, nor will the experience be the same for everyone. Simone de Beauvoir suggests that the human body is ambiguous; subject to natural laws and to the human production of meaning,

> It is not merely as a body, but rather as a body subject to taboos, to laws, that the subject becomes conscious of himself and attains fulfilment – it is with reference to certain values that he valorizes himself. To repeat once more: physiology cannot ground any val-ues; rather, the facts of biology take on the values that the existent bestows upon them.
>
> (de Beauvoir, 1989: 76)

Bodies as represented, for example, as marginalized, also experience themselves and are crucial to an understanding of selfhood and the processes through which people position themselves and are positioned within the social world:

> the body is not a thing, it is a *situation* ... it is the instrument of our grasp upon the world, a limiting factor for projects.
>
> (de Beauvoir, 1989: 66)

This approach provides a way of bringing together the natural, mate-rial body, the experiences of embodied selves and the situations, which include representations, practices and policies, which recreate the lived body. Bodies are not 'just' in a situation, nor are they just objects of empirical enquiry; bodies are more than this. De Beauvoir's analysis of the 'lived body' provides a means of enabling

> a situated way of seeing the subject based on the understanding that the most important location or situation is the roots of the subject in the spatial frame of the body.
>
> (Braidotti, 1994: 161)

Bodies can be situated on the margins through structural factors such as economic inequalities, racialization, ethnicization, discrimination of grounds of gender and of physical or mental impairment, but bodies are also themselves situations through which people experience themselves, both negatively and positively. As Toril Moi argues, 'To claim that the body *is* a situation is not the same as to say that it is placed *within* some other situation. The body is both a situation and is placed within other situations' (Moi, 1999: 65).

Embodied selves, understood through the trope of lived bodies, accord greater agency and possibility for transformation and avoid the reduction of the self to the body by acknowledging both the situations which bodies inhabit and the interrelationship between bodies and situations: bodies are situated and are themselves situations. De Beauvoir argues that to claim that the body is a situation is to acknowledge that, for example, having a woman's body is bound up with the exercise of freedom. The body-in-the-world is in an intentional relationship with the world, although, as Young argues, women do often end up living their bodies as things (Young, 2005b). Lived embodiment disrupts dichotomies of mind and body, nature and culture, public and private and foregrounds experience (Young, 2005). If embodied selves are targeted in the interventions in sport which aim to bring the margins into the centre, seeing bodies as situations might accommodate materialities and permit understanding of how sport can be the site of political activism and agency. There are the connections too between the bodies that are targeted and those do the targeting.

Young's feminist phenomenological approach, which deploys the concept of embodiment, attempts to redress the imbalance in Merleau-Ponty's work by focusing on gender and in particular the specificities of women's embodiment. Young challenges the universal account of the gender neutral body implied by Merleau-Ponty and claims that the female body is not simply experienced as a direct communication with the active self, but it is also experienced as an object. She suggests that there are distinctive manners of comportment and movement that are associated with women. Young attributes these different modalities; first, to the social spaces in which women learn to comport themselves. In terms of sport this involves constraints of space and repeatedly acting in less assertive and aggressive ways than men. Conversely, from this it might be deduced that men acquire those embodied practices, as in boxing, which are aggressive. Second, Young suggests women are encouraged to see themselves through the gaze of others including the 'male gaze', as developed in the work of Laura Mulvey (1975), and to become more aware of themselves as objects of the scrutiny of others. The aspirations to the heroic body of the successful boxer might be viewed as informing the dreams and the practices of men who box. Whereas, young women practise the comportment of femininity, young men engage in the techniques of masculinity, embodied in the 'hard man' image of the boxer. Social space is constituted by body practices and culture in which psychic investments are made; the gym is not a world apart. Within gender binaries, not only are women expected to be less physically aggressive, but also their anxieties might assume corporeal expression. This could be in the form of Lafferty and McKay's 'soft boxing', which is the equivalent of 'throwing like a girl', or covering their eyes during scenes of violence, whether in the cinema or on screen, or, in the case of boxing, at a fight. Spectatorship too is gendered and being able to watch cinematic scenes of violence is also constitutive of some versions of masculinity (Woodward, 2007b, 2008b).

Theoretical explanations which address gender differences provide a means of explaining some of the dynamics of embodied experience in sport and especially of bodies on the margins, which often include gender, although much of the political activism on diversity has concentrated on race, as Chapter 4 demonstrated.

Diversity politics

As was argued in Chapter 4, the politics of diversity has seeped into sport as it has in other areas of social life. It has become ordinary in many situations such that it has become possible for the margins to be reconstituted within the terrain of popular culture. It is not that there are no longer 'racial events' (Doane, 2006), or that racism is not routine, but bodies on the margins have been resituated, albeit often in incremental ways. Policy interventions to promote social inclusion have two interrelated aspects; the role played by regulatory bodies, including sports clubs, and that of 'the grass roots campaigns to save clubs' both concerned to 'mobilize prevailing interpretations of the word "community"' (Wagg, 2004: 16). Community is an avowedly complex concept, as acknowledged by both academics and practitioners (Brown et al., 2006). It is the relationship between 'outsider' communities, situated as disadvantaged, and the related notion of belonging and 'insiders', in which people situate themselves. For example, in the case of football, where club allegiances and affiliations have been strongly linked to particular 'local' clubs, many of which have become increasingly global, especially in their recruitment of players, but also in terms of fan-base, social change has led to reconstituted allegiances of class and kinship and in the geographical location of communities and economic factors have led to the exclusion of some groups in society and thirdly, the notion of 'community has been distorted and undermined because it has been "sprayed on" to all manner of initiatives to indicate feelings of inclusiveness and the overcoming of social deprivation' (Football Foundation, 2006: 9).

'Community', as deployed in sport, is described as embracing 'particular groups such as current and potential supporters from black and minority ethnic "communities", those with disabilities and those from "disadvantaged" groups that are considered as such...those who have been labelled "community" targets' (Brown et al., 2006: 49). Women and girls are seen as belonging to a community in this sense because they too are bodies situated on the margins of sport where their physical absence has been marked either as practitioners or as supporters at football grounds, although the sport is increasing in popularity. For example, the English Football Association's enthusiasm about its own commitment to promoting the women's game (FA Women, 2007) is not matched by the visibility or embodied presence of women (Caudwell, 2003). This is illustrated by the negative situating of women. Women's teams are the first to be sacrificed when economies have to be made, even by clubs with an exemplary record in promoting diversity (Charlton Athletic, 2007) as in 2007, when the UK football club, Charlton Athletic, saved £250,000 by disbanding their women's team (Leighton, 2007).

Fans play a significant part in the promotion of practices of diversity and social inclusion, historically through the development of organizations like Kick it Out (KIO) and Football Against Racism in Europe (FARE) and through the routine, embodied practices with which they engage. Organizations which challenge social exclusion often do so within the framework of human rights, through the articulation and promotion of anti-racist practices and strategies, are implicated in negotiating new meanings for bodies on the margins and actively engage in 'explicit and systematic encounters with prevailing philosophies of race and ethnicity' (Schwarz, 1996: 81). Racialization and ethnicization not only situate embodied selves on the margins, but they also constitute the lived bodies as situations. Anti-racism and discourses of social inclusion challenge and seek to subvert these philosophies of race which exploit different ideas about nature and culture and the naturalization of embodied differences. Most exclusionary apparatuses are imbricated in the embodied practices of sport, including ideas about 'race', gender and able-bodiedness that are embedded in a naturalization of 'race' (Wade, 2002) and the assertion of heterosexuality, as Jayne Caudwell argues using the example of women's football (2003), and draw upon repertoires of what is deemed 'natural'. This naturalization is more explicit in relation to gender through the segregation by regulatory bodies of sport into the women's and men's games (Hargreaves, 1994). Football, like most sports, is highly gendered. The regulatory bodies which discipline and organize football are clearly segregated, although it is only the women's game which is marked as such. The 'World Cup' is the men's competition and gender is largely only noted when women are playing.

Routine encounters and body practices as well as more dramatic events and campaigns such as FARE's events and Street Kick organized at the 2006 Men's Football World Cup create new ways of situating embodied selves (Woodward, 2009) Whilst some fans may be the motor for change, others, through their physical presence and body practices, are a source of social exclusion, which prevents transformations in the embodied identities that can be taken up in sport. Racism has, nonetheless, been explicitly addressed through attempts to attend to those bodies on the margins; conviviality, despite its name, is both negative and positive and usefully describes the ambiguities and tensions both between the positive and negative dimensions of the relationship between the anti-racist bodies and the embodied selves with whom they engage and the disjuncture between policies and practices and especially the lived, situated experience of those perceived as in the centre and those on the margins. Although combating racism is not and never has been central to sport, it has, however, become a material possibility, so that situations can be reconstituted and bodies resituated in sport.

Lived bodies open up opportunities for new ways of belonging that transgress traditional limitations that are spatially, corporeally and temporally located, albeit often in uncomfortable ways; progress is limited and contradictory. However, change is explicitly acknowledged by activists, albeit within the context of resilient racist discourses, but transformations and new identifications are apparent in the increased visibility of a more diverse constituency at grounds and on the websites. New fans and activists may further challenge the distinctions that have been between 'authentic' and 'inauthentic' supporters and exercise an agency that counters the culture of dependency and the notion of being victims in the classification of under-represented groups in sport.

Conclusion

Phenomenological accounts offer a shift of focus from the discursive field of governmentality and policy-making at the level of governance to the lived experience of embodied selves through the concept of the lived body. Developments of phenomenological concepts and those of body techniques permit an analysis of what is done to the body and what the body does (Crossley, 2001) or, as de Beauvoir expressed it, the situated body and the body as situations are formulations that allow for the possibility of action and agency and distance a theory of embodiment from naturalization and biological reductionism or even corporeal determinism.

There is no simple, linear trajectory of progress and change, but a series of disruptions and realignments which do, however, suggest that new identifications are emerging, even in sport which is so marked by binaries and inequalities. It is only possible to theorize change if there is an acknowledgement of the possibility of creativity and collective action. Whilst dominant hierarchies persist, as de

Beauvoir argues, constraints can be not only structural as in the operation of patriarchal and colonialist discourses of exclusion, but also corporeal in terms of the impairment and disease that flesh is heir to. However, an understanding of embodiment that challenges distinctions between body and mind and takes on the political project of transformation has to incorporate the dynamic of an embodied being-in-the-world. Feminist phenomenology offers an entry point into linking the structural constraints of the wider social and cultural arena with the body as situation with which they are inextricably linked in a version of 'we are our bodies' that permits the politics of action and activism.

By focusing on situated embodiment and bodies as situations it is possible to note the ambiguities of bodies in the policies aimed to promote social inclusion and in embodied sporting practices. The experience of social exclusion encompasses all the aspects such as gender, 'race', ethnicity, sexuality and dis/ability that are increasingly the focus of the policies and practices of organizations promoting cohesion. Those who challenge social exclusion have to negotiate their own situations. Bodies mark the visibility of differences and inequalities as well as the triumphs which human beings are capable of achieving.

De Beauvoir's feminist phenomenological understanding of bodies as situations as well as situated by external circumstances combines regulating bodies, for example, through the policies and practices through which governments attempt to recruit citizens and material embodied practices and allow for some understanding of the ambiguities and contradictions of experience. This is an alternative social constructionist reading which enables change, albeit within the context of constraint, social limitations and ambiguities. Phenomenology is not only a methodology that creates space for everyday practices and allows the voices of the subjects of research to speak. The feminist versions discussed in this chapter also provide a theoretical framework within which to situate enfleshed bodies and regulatory practices. Thus it is possible to challenge some of the binary logic which is so entrenched within sport, especially in relation to mind and body and what is natural and what is socially constructed.

This conceptual framework focuses more upon the embodied practices of practitioners than post-Foucauldian interpretations of regulatory mechanisms that target the body. Some phenomenological accounts within sport have concentrated on body practices than on disciplinary mechanisms, for example of regulatory bodies, but the concept of the body as situation and as situated permits some understanding of all aspects of embodiment in sport, including spectatorship, although this has not been central to phenomenological accounts or those based on body techniques, such as Wacquant's, which have sought to demonstrate that sport is about what athletes do more than how sport is represented; body practices rather than cultural construction. Sport is both. In order to understand the mechanisms and practices through which sport is experienced and constituted, it is necessary to combine all aspects of sporting body practices and sporting bodies and the media, however sensational and prone to hyperbole, are a key component aspect of the regulatory bodies that shape sport. The interrelationship between representation and materiality is an aspect of dualistic thinking that phenomenological theories engage with less directly. Sporting embodied practices include spectatorship, which is the particular focus of the Chapter 6, which moves into the terrain of sensation as well as the sensational by exploring the ways in which sport is represented and experienced through spectacle and spectatorship. Embodied sporting practices include sensations that are experienced in spectatorship and different dimensions of corporeal presence in all aspects of sport. By focusing upon embodiment in all its dimensions, it is also possible to highlight difference at all levels of sporting experience.

6
Beyond Text: Spectacles, Sensations and Affects

Sport invites spectatorship and very powerful allegiances, for example, through the identifications of fans and supporters that go way beyond the practices that are being enacted in its performances. The team might even stand for the nation at some moments and the history of sport is replete with dramatic and iconic moments. Sport elicits and creates passionate feelings through the enfleshed experiences of participants and of spectators who, through the senses, are drawn into the exhilaration of athletic achievement even vicariously. Sports fans play a key role in the co-production of spectacle and in the spectacular (Gamson, 1994; Rojek, 2001). Fandom too is constituted through the regulatory bodies of sport. This can be illustrated by the different experiences of British football and the Football Association, when contrasted with the US NFL. The Premiership, and indeed the Football League, is dominated by the top four or five clubs. Other clubs, especially those outside the Premiership, or only recently promoted to it, struggle, often to the point of facing administration and bankruptcy. The Super Bowl is one of the most spectacular mega events in the sporting calendar, but NFL regulations mean that this event always carries the democratic potential of including different clubs in the league. The NFL does not permit the massive transfer deals and huge salaries of the British Premiership and thus promotes a more egalitarian game in which fans from all clubs have a chance of making it to the Super Bowl.

Sport is all about the 'quest for excitement' (Elias and Dunning, 1986), although not necessarily as part of the civilizing process. As Elias and Dunning (1986; Dunning, 1999) recognize, however, sport can also be violent and no matter how sport is regulated and disciplined it has the power to generate passions in which bodies and senses are central that are violent as well as dramatic. The emotions elicited by sport and sporting encounters also indicate the darker side of passion, irrationality and the disruptive effects of emotion, which suggest the operation of unconscious as well as conscious forces. Sport offers opportunities for particular versions of sensate experiences lived in and through the body that carry risks and dangers of disruption as well as the disciplinary regimes of control and cohesion. The pleasures of sport can also, of course, be more subtle, as expressed in the slow nurturing interactions of cricket, which offers more complex forms of enjoyment than the immediacy of soccer or baseball (Brearley, 2009). The pleasures of sport are diverse and encompass complex satisfactions, thrills and excitement and risks and dangers.

Meanings about body practices and embodied selves are recreated very powerfully through the public arena of representation and the media as well as routine practices of training in the gym, which have been central to phenomenological accounts. The media sport nexus involves a symbiotic relationship between sport, commerce and the media where the sponsorship upon which sport depends in the globalized networks of contemporary life is also largely dependent on the massive media coverage which sport achieves (Crawford, 2004; Gratton and Taylor, 2004). Celebrities, whether configured as sporting heroes or villains, play a key role in the media sport nexus and are a key component in the production of powerfully expressed and often deeply felt meanings about embodied sporting identities. Previous chapters have focused on those who are underrepresented or marginalized within sport; this chapter explores the status of representations and the identifications that are made by examining the phenomenon of the sporting spectacle and the celebrities who inhabit these spaces. Celebrities constitute an increasingly important aspect of sport (Turner, 2004). Sports celebrities dominate the media. This is not to say that sport has not always had well-known figures; Babe Ruth, Jack Dempsey, Donald Bradman, Stanley Matthews, Freddie Truman, George Best, Muhammad Ali (Cashmore, 2005). Cashmore's list is all male, which is hardly surprising given the visible presence of male athletes represented as heroes and the relative absence of women so configured. This is changing, but very slowly and women are more present as sexual partners of male athletes, for example as the 'wives and girl friends' or WAGs so beloved of the popular media and celebrity magazines, which make up a genre devoted to the very idea of celebrity and where pop music, soap opera and the cult of 'personality' elide with sport. Not all male celebrity bodies conform to stereotypes of heroic masculinity, however, or at least, this heroic masculinity is configured differently in particular sports across the globe. Those sports which offer more subtle pleasures also give priority to aspects of embodiment that privilege skill and talent above conformity to contemporary standards of physical beauty. Heroic figures of Indian cricket like batsmen Sachin Tendulkar and Sunil Gavaskar embody a spiritually inflected cricketing aesthetic, that is far more significant than more conventional signifiers of hegemonic muscular masculinity (Brearley, 2009).

Much of the discussion about the configuration of sporting heroes and the role of the media in their construction has been based on sensationalized narratives of success, sometimes offset by failure, but framed by the trajectories of a biographical story. Sports stories vie with other narratives in media coverage of popular culture. Narrative presents one way of understanding how sport works and why it has such a hold on collective imaginations and belongings. Stories which create heroic figures as their central protagonist have a long history and can take an immediate hold on the collective imaginary through the cumulative logic of narrative, which also underpins some phenomenological accounts.

This chapter also suggests some alternative understandings of the making of celebrities, the media and sensation in sport, by exploring the idea of affect. More recent conceptualization of affect has developed out of, and in contrast to, psychoanalytic approaches in order to address some of the tensions between representations and material body practices. Emphasis has been placed upon the need to investigate some of the processes that are implicated in the intersection of representation and the materialities of embodiment in order to question the discursive production of meaning by examining the material, embodied, speaking, feeling subject (Irigaray, 1991).

The interrelationship between the corporeal and the discursive is well illustrated in sport by media representations of live events, for example through television and film and there are distinctions that can be made, especially between the live event and its representation on screen (Oates, 1987), for example in terms of sensation(s) and the experiences of the senses. Film, however, can approach the intensity and has been used to recreate powerful feelings of affinity and identification. Film can be explicitly political as in Leni Riefenstahl's filming of the 1936 Olympics, which followed her *Triumph of the Will*, released in 1934 as explicit propaganda for Hitler's Nazi party. Sport has a place in cinematic history for its evocation of enfleshed sensation and experience; Martin Scorsese's *Raging Bull* (1980) is a particularly good example of the technically superb appropriation of sport as a vehicle for the exploration of affect. The interrelationship between the discursive and the corporeal is used in this chapter to explore the assemblage of sensations which constitute the embodied self (Sobchack, 2004) to move beyond visibility and the visual to include aspects of the sensate and sentient body and the role of feelings and affect in the field of sport. This chapter also traces some of the histories of the sport media in order to map some of the intersections of sport and media technologies, by shifting the emphasis onto more public representations of sport and explores the phenomena of spectacle, celebrity and heroism as key components of the interface between the discursive and the embodied. What is spectacular about sport?

Spectacles and the spectacular: Invoking sensation

The spectacle, as an imposing public display, is not simply a benign cultural form, because it cannot be separated from 'non-coercive strategies of power and persuasion' (Cary, 2005). High-profile sporting events staged for the entertainment of the public, like the NFL Super Bowl, Olympic Games or FIFA World Cup, which, through their celebratory sporting veneer, routinely communicate the values and ideologies of the dominant corporate global capital. Spectacles are assemblages of systems of power, multiplicities of idealist cultural aspirations, amateurism, professionalism, commercial interests and material bodies. The sport spectacle has, of course, long been a vehicle for the expression and/or performance of dominant cultural practices and sensibilities. This can be attributed to the visceral intensity of competition-based physical culture that possesses an almost unrivalled capacity to capture the interest and imagination of publics located within divergent historical and social contexts. The embodied practice of being a spectator at displays of competitive physicality has played an important role in establishing the socio-cultural import of some physical activities over others. Thus, the Olympic festivals of ancient Greece, the gladiatorial contests of ancient Rome, and even the folk football rituals of pre-industrial Europe were only fully constituted as socially, culturally and, indeed, politically significant practices through the presence and involvement of huge numbers of spectators, who also constitute this sporting assemblage.

Sporting spectacles are focal points for popular identities, desires and fears. The circuses of violent entertainment staged within Rome's vast amphitheatres were a means of appeasing and creating the baser sensibilities of the Roman populace as well as providing a diversion for the disenfranchised. These games constituted a highly visible site for the elite to exhibit their economic and political power, for example, by sponsoring extravagant, bloodthirsty spectacles, including gladiatorial combat, elaborate and voluminous human sacrifices, the mass slaughter of animals, and even carefully staged naval battles all as part of generating the popular approval of the masses, but haunted by the threat of violence.

Modern spectator sport from the late nineteenth and early twentieth centuries incorporated considerably less savage forms of competitive physicality than their ancient antecedents; they did nonetheless perform similar socio-political functions. The instantiation of modern sport during the nineteenth century both disciplined and commodified popular physical culture through the standardized regulations and reconfigured systems which provided opportunities for aspirant participants, entrepreneurs and spectators, which were taken up fully in the twentieth century.

One of the mega events which, although amateur in its founding principles, has played a key role in the development of sporting spectacles is the Olympics, which illustrate the media sport nexus (Miller et al., 2001; Tomlinson, 2007). Such mega events are large-scale cultural moments, which combine dramatic character, mass popular appeal and international significance (Roche, 2006) and which also have the power to make some things and some people visible. Visibility, even at the supposedly democratic Olympic Games, has been problematic in the case of black and minority ethnic competitors and women, who were excluded from the first Games in the modern period. The most explicit and visible exclusion and attempt at invisibility was the 1936 Berlin Games, which were the Games most directly engaged with promoting a political ideology; Hitler found the success of black athlete Jesse Owens particularly troubling. Sporting spectacles provide spaces in which transformations can be represented although their very spectacular properties create the danger of change being translated as tokenism. Cathy Freeman, the Aboriginal sprinter's visibility in 2000 at the Sydney Games, was marked by her carrying the torch at the opening ceremony and she won Australia's one track gold medal, but spectacles offer the danger of tokenism and her presence at the mega event did not coincide with material change in the lives of Aboriginal people.

The global spectacle of the Olympic Games was expedited through successive phases of technological advancement in the mass communications industry that allowed the Olympic spectacle to engage and inform, both internal and external audiences (Guttmann, 2002). Initially advanced as a festive celebration of sporting excellence, fair play, amateurism and internationalism, the modern Olympic Games soon became transformed by the intersection of different power axes, although there has been some consistency in the dominance of white male elites on the IOC. Different axes of power have come into play along the lines of economic, social and political systems. Thus, the Games have become forums for the international display and attempted validation of specific political-economic orders, such as those associated with British imperialism, London, 1908, German fascism and the supposed celebration of Aryan bodies, Berlin, 1936, postwar regeneration and optimism in London in 1948, USSR communism, Moscow, 1980, US capitalism, Los Angeles, 1984, and US neo-imperialism, Salt Lake City, 2002, and the advance of neo-liberalism into former communist states, Beijing, 2008, with the most spectacular opening and closing ceremonies in the history of the Games. Many recent Olympic celebrations, apart from Salt Lake City, 2002, the Games have been made spectacular, particularly through the performance of defining national cultural characteristics within Olympic opening ceremonies, and the utilization of specific national geographies as event locations and facilities (Tomlinson, 1996). In this manner, Barcelona, 1992, Sydney, 2000, Athens, 2004, and even Beijing, 2008, are all illustrative of the commercial processes through which the Olympic Games have become implicated within, and veritable motors of, what are the over-determining forces and networks of global networks of consumption.

In the second half of the twentieth century sport was conclusively and apparently irreversibly integrated into the commercial ferment of market economies and consumption. Many sporting bodies, such as Major League Baseball (MLB), the National Basketball Association (NBA), the National Football League (NFL) and the National Hockey League (NHL), originated as professional, putatively commercial, ventures, but by the mid-twentieth century sporting bodies were fully commercialized. Sporting bodies became incorporated into the assemblages of the wider culture (Miller et al., 2001).

The demands of understanding the spectacular economy of contemporary sport culture lead, somewhat predictably, to Debord's (1990, 1994) theorizing on the society of the spectacle. However, as Tomlinson (2002) has argued, Debord's provocative treatise on the transformations in relations between capitalism, technology and everyday life are the subject of superficial invocation as if sporting themselves encapsulated the complexities of spectacular society. The tendency towards reifying the spectacle can be deconstructed through recourse to Debord's theses which unravel the layered complexity and multi-dimensionality of the spectacle, and its position and function within spectacular society. Such critiques of spectacle and the spectacular assume the success of the processes through which spectacles work. The systems of such analyses largely underestimate the impact of the spectacle upon the spectator, although when there is disruption this may provoke some attempt to interrogate some of the mechanisms and embodied experiences of spectatorship.

The real thing: Spectatorship

Sport is often sensational and, in the public arena, sensationalized, for example, through the emotional responses of spectators that can take extreme forms, although the rhythms of the stands and terraces routinely involve the expression of powerful feelings and discord. However, sensation is also part of more everyday sporting spectatorship. Boxing offers a particularly dramatic example of the routine invocation of sensation. Joyce Carol Oates argues that boxing is not drama but reality, because she suggests that, first, boxing is about flesh and blood and real, intended damage and, second, actual physical combat and being present at the ringside is authentic whereas cinema or television spectatorship is sanitized and second hand (Oates, 1987). There are, of course, multiple ways of defining what is real, one of which is to appeal to the Lacanian 'real' in order to address that which is otherwise inexplicable.

The Lacanian 'real' might be theorized as more than embodiment and in excess of that which is produced through a discursive outside. Whilst discursive approaches like Foucault's focus on how people are positioned within discourse, Slavoj Žižek, conversely, develops the Lacanian distinction between the real and the reality and argues that the real is outside discourse and cannot be accommodated or described within it. Žižek and Dolar argue, 'the ultimate lesson of psychoanalysis that human life is never "just life": Humans are not simply alive but are possessed by the strange desire to enjoy life to excess, passionately attached to a surplus that derails ordinary run of things' (2002: 107).

Sport offers moments of excess as that which cannot be symbolized. It also presents what Stallybrass and White call the 'attraction of repulsion' (Stallybrass and White, 1986: 140) which might also constitute excess through the congruence of that which is appealing and that which is disgusting and repugnant, whether of sensitivity to aggressive, hostile behaviour or the repulsion of physical injury. Embodied sporting practices lead to injuries and damage. Sport is not only about beautiful, healthy, fit bodies, even those of its celebrities. Celebrity sports' stars break down and fail; this can be on the field or in the ring, as in the case of boxing tragedies and disasters (Michael Watson and Gerald McClellan), or in the ring and out, as in the case of Mike Tyson. Even Olympic medallists can find holding onto heroic status difficult outside their sport, although these might be minor aberrations as in the case of swimming star Michael Phelps's experimentations, or more serious one; other athletes find pharmaceutical assistance hard to resist like sprinter Ben Johnson, whose offences were detected. Others' tragic early deaths, if not officially recorded as drug-induced, like Florence Griffiths Joiner, are part of these histories. Athletes are well known for being well known but their attraction may not be so appealing and sport has its examples of excess that are anything but attractive.

However, as Stephen Frosh points out, even among those who support the usefulness of a Lacanian 'real' trying to write about what is not expressible is difficult without lapsing into mysticism (2002: 152). However, the other aspect of writing about symbolizing that which cannot be symbolized is that where it is attempted it is more often done in the context of the avant-garde (Kristeva, 1987) or work of some artistic merit (Žižek and Dolar, 2002) than in relation to popular culture. However, this is not to say that the extra discursive cannot be accessed through more popular media, including sport.

Žižek's Lacanian 'real' is more than a Freudian unconscious. This is a real which is experienced as an absence, but nevertheless an absence which is ubiquitous and dogs our every track. The real is that from which we recoil, illustrated by Sigourney Weaver in *Alien* recoiling from the monstrous figure of the alien, 'the subject constitutes itself by rejecting the slimy substance of *jouisssance*' (1993: 62). This is a very different *jouissance* from Irigaray's. In some of his work Žižek (1992) presents the real as disgusting; a monstrous entity, which can also be linked to pleasure, but only insofar as it involves the inseparability of revulsion and titillation; looking at that which is disgusting or humiliating, like turning to look at road accidents or watching violent films or watching violent sport the attraction of which is premised on damage. Bare knuckle fighting as presented in David Fincher's *Fight Club* (1999) or in the illegal activities that sometimes operate on the periphery of boxing gyms (Beattie, 1996) or the recent resurgence of interest in cage fighting and less regulated forms of boxing are part of these processes. The real exists only in contradistinction to reality and thus matches the limitations of languages. Sarah Kay suggests the Lacanian trope of describing the real as something which is 'stuck to the sole of your shoe and which cannot be removed' (2003: 4). Žižek's real seems more useful in locating disgust and alienation than pleasure, although its ambiguity, which must surely be necessary because it cannot be symbolized, affords a point of entry into that which cannot be explained discursively or which presents disruption to the shared norms of performance.

The real offers one route into the instatement of the psychic into the discursive which Judith Butler has agreed is a necessary antidote to Foucauldian versions of the social (1993) in her engagement with synthesizing the discursive with the psychoanalytic in a theoretical framework which redresses the excesses of discursive social construction. However, she forcefully rejects the real upon which Žižek's whole explanatory framework is based because of its deeply misogynist relationship to gender difference, which is a crucial stumbling block within psychoanalytic theory; acknowledging difference and the centrality of gender and exploring its mechanisms are different from asserting the inevitability of particular versions of gender supremacy and in particular patriarchy. However, there are some points of consonance which might make Žižek's real redeemable or even applicable. Butler agrees that any 'theory of discursive constitution of the subject must take into account the domain of foreclosure, of what must be repudiated for the subject itself to emerge' (1993: 190) but she too points to the problematic of symbolizing what cannot be symbolized, although she accepts the usefulness of the phantasmatic promise of political identification which illustrates Žižek's reworking of the Althusserian understanding of interpellation, which posits that identifications occur at the level of the unconscious through a process of hailing or naming.

An alternative configuration of the real is presented by Vivian Sobchack, in exploring some of the tensions between fiction and documentary within film. She suggests that the 'not real' and the *irreal* are not the same thing (2004). Whereas the 'not real' covers that which is fantastical and impossible or even implausible fiction, we know what is 'not real' because we know what is real. What is 'not real' could not exist in the real world against which its veracity or validity is assessed. The *irreal* on the other hand is not judged against the real; the *irreal* is the imaginary. The *irreal* involves a suspension of belief which means putting aside what we know to be real. Sobchack's discussion is of embodied knowledge and cinematic consciousness and she is using the devices of film, including documentary footage, cartoon animation and the deployment of embodied actors but her distinction between real and *irreal* has some application to the specificity of sports spectatorship through its diverse mediated forms. This conceptualization provides an alternative to the categories of real, meaning true or authentic, and the antithesis, false, not true or inauthentic, as measured against the veracity of what is real. Sport is mediated in these ways and the third category of the *irreal* presents an alternative to the real of the live performance and the unreal or inauthentic of representation; a binary that permeates sport in terms of true, authentic fans who are present at every game, as opposed to inauthentic fans or fair weather supporters, who watch on TV.

The *irreal* also offers a route into engaging with the corporeality of spectatorship; watching sport, whether live or mediated, is physical and engages the senses as well as playing sport. Live sport or even mediated sport on the television or radio is physical for spectators; they are drawn into the spectacle and, as Vivien Sobchack argues, spectatorship is a corporeal practice in which all the senses are implicated (2004). Embodiment is central to spectatorship, through the visibility of bodies on the field or screen and the primacy of the visual through the medium which demonstrates powerfully the 'embodied and radically material nature of human existence' (Sobchack, 2004: 1).

In the field of sport, the researcher too is physically situated, although, if not a participant, for example in the gym as many ethnographers have been in a sport such as boxing, at more of a distance than the spectator. For example in boxing, although in less dramatic an experience, the researcher in the gym witnesses the regimen of physical exertion in sparring or bag work, which nonetheless can be powerful. You can smell, hear and see the experience. At a fight this is enhanced and the spectators' reactions are also physical sensations. Boxing reminds us of the flesh at every turn. Joyce Carol Oates suggests this is embedded in the genealogy of boxing:

> When the boxing fan shouts 'Kill him! Kill him!' he is betraying no peculiar individual pathology or quirk but asserting his common humanity and his kinship, however distant, with thousands upon thousands of spectators who crowded into the Roman amphitheatres to see gladiators fight to the death. That such contests for mass amusement endured not for a few years or even decades but for centuries should arrest our attention.
>
> (Oates, 1987: 42)

Oates is here attempting to combine empirical observation of the primordial, visceral passion expressed at fights with an explanation that locates such feelings in the genealogy of collective affect. Invoking the history of such raw emotion and its ancient comparable expression might lend weight to the depth of feeling, however unpleasant, and, in a sense, naturalize it. Boxing, with its one-on-one corporeal combat and intentions of injury to one's opponent, is more extreme in exciting such emotion, but many sports, especially team sports like baseball and football, with powerful identifications elicit similar expressions from the crowd. Locating the sensations of smell, sound and sight suggests a phenomenological focus on the lived body and an analysis of sensation as the experience of the senses which is experienced by the embodied self who so describes it (Sobchack, 2004). This differs from psychoanalytic accounts, which suggest the operation of an unconscious which can be accessed and work according to a set of laws that are distinct from those of the conscious mind. This account of embodied sensation draws on the lived body experience of phenomenology which also includes both reflection of the situation of the researcher and a methodology that engages with sensation as part of being in the world and a framework of narrative, presenting a sense of order and making sense through sequencing.

The Deleuzian conceptualization of sensation takes a very different approach; sensation is a two-way process which does not highlight the drama of sensation or essence in its vitalist account. This is linked to the concept of assemblage. Sport constitutes a system which brings together the body practices of engagement and the representations and those who so engage and the discursive field of comment. In this sense sport is an assemblage which 'in its multiplicity necessarily acts on semiotic flows, material flows and social flows simultaneously' (Deleuze and Guttari, 1987: 22). An assemblage establishes connections with previously separate entities reuniting them into a complex whole. This emphasis on mobility challenges the fixity of oppositions, which are nonetheless deeply embedded in the discourse of sport and in the voices of its participants. Everything is mixed up in assemblages that suggest an interaction and intertwining of forces, not the determination of one force by another. This has resonance for schemata which suggest a single cause or a simple relationship. Sport provides assemblages that include the articulation of the corporeal and discursive elements of a given assemblage. For Deleuze the elements of assemblage change and when they do so does practice and experience, which also allows some possibility of productive transformation (Deleuze and Guttari, 1987).

Sensation and affect

In order to explore some of the ways in which the concept of affect can be deployed to explain some of the ways in which sport and especially sporting spectacles and heroic figures might be central to understanding how sport works and in particular how it is able to generate such passions, it is necessary to pay some attention to what has been termed the affective turn (Clough, 2007). The engagement with affect has drawn upon the work of a range of thinkers, many of whom have been developing some of the ideas of Deleuze and Guttari (1987). Although Deleuze argues for a material understanding of becoming that is not only radical in the challenges it offers to existing philosophies but also complex in its treatment of bodies, the approach has considerable purchase for addressing the problem of bodies and of sensation and affect. For my purposes in this chapter it is Deleuze's focus on the contact points and mergers between representation and materiality and his rethinking of the mind body relationship in the context of affect that is particularly pertinent. Bodies are not central to Deleuze's thought and his materiality is particularly fluid. However, as Rosi Braidotti points out in Deleuzian theory, bodies are the site at which different forces are in operation, including the corporeal and the symbolic.

> The 'body' as theoretical topos is an attempt to overcome the classical mind-body dualism of Cartesian origins, in order to think anew about the structure of the thinking subject. The body is then the interface, a threshold, a field of intersecting material and symbolic forces.
>
> (Braidotti, 1994: 169)

This establishes my concern here with the impact of ways in which sport is mediated and represented and the relationship between the embodied selves that participate in sport and that are represented and the modes adopted by the media in creating meanings about these sporting bodies and sporting selves. This takes place within a culture in which sensation and the sensory dominate.

> Transformations of our sensory, perceptive and conceptual habits are everywhere in contemporary culture and because they are everywhere, they have become invisible. They have permeated the social and cultural spaces bringing about a collective fascination with speed, acceleration and intensities. Mood-enhancement is another word for cultural consumption.
>
> (Braidotti, 2002: 269)

Sport provides spectacles of physical and psychic daring which both challenge and attract the sensory apparatus of the spectators, leaving spectators hungry for more. At its best sport can provide a literally sensational experience for both the participants whose exhilaration can reach levels of intoxication that de-materialize the body and the spectators whose enhanced experience overwhelms traditional modes of perception. Social and cultural norms act as magnets which draw the subject into the spectacle and the corporeal experience.

Comparing sensation in the context of art and drawing particularly on a critique of the French Impressionist Cezanne as illustrative of his argument, in his book on the painter Francis Bacon, Deleuze describes sensation as

> the opposite of the facile and the ready-made, the cliché, but also of the 'sensational', the spontaneous etc. Sensation has one face turned towards the subject the nervous system, vital movement, 'instinct', 'temperament' – a whole vocabulary common to both Naturalism and Cezanne) and one face turned toward the object (the 'fact', the place, the event). Or rather, it has no faces at all, it is both things indissolubly, it is being-in the – World, as the phenomenologists say; at one and the same time I *become* in the sensation and something happens through the sensation, one through the other, one in the other.
>
> (Deleuze, 2005: 25)

Thus sensation is part of the becoming process which characterizes Deleuze's approach. Similarly, affect is often distinguished as that which is sensed rather than known, an impersonal force underlying processes of becoming, and as such provides the potential for a distinct site of enquiry. There is considerable slippage in the use of the term and overlap, in particular, with feeling and emotion, but affect is distinguished although not entirely separate and distinct from these terms. For example, where

> *feeling* is often used to refer to phenomenological or subjective experiences, *affect* is often taken to refer to a force or intensity that can belie the movement of the subject who is always in a process of becoming ... Although affects might traverse individual subjects, for many scholars they undo the notion of a singular or

> sovereign subject...Affects do not refer to a 'thing' or substance, but rather to the processes that produce bodies as always open to others, human and non-human, and as unfinished rather than stable entities.
>
> (Blackman et al., 2007: 6)

The concept of affect is seen as asserting part of the vitalist turn in social theory which is a useful means of redressing the imbalance that has been perceived that Blackman has suggested has led to an overemphasis on the social and discursive at the expense of understanding feelings, emotions and sensation (Blackman, 2007, 2008a, 2008b).

Patricia Ticineto Clough suggests that there has been an expansion of transdisciplinary and interdisciplinary interest in affect which constitutes an 'affective turn', which has to be located within a contemporary world marked by crises of terrorism, war and torture that demand some explanation from critical theorists. As she argues,

> Affect refers generally to bodily capacities to affect and be affected or the augmentation or diminution of a body's capacity to act, to engage, and to connect, such that autoaffection is linked to the self-feeling of being alive – that is aliveness or vitality.
>
> (2007: 3)

Affect is not only understood in terms of the human body; it is also conceptualized in relation to those technologies that make possible the visualization of affect and to produce affective bodily capacities that go beyond the physiological constraints of the material body. Thus for those who espouse the transdisciplinary interest in what Clough, following Brian Massumi (2002) calls the affective turn, expresses a new configuration of bodies, technology and matter that are necessary to an understanding of the social. Although this conceptualization of affect is self-reflexive, affect challenges different dualisms, including mind and body and rationality and passion or emotion. Affect encompasses the inextricable links between the power people have to affect the world and the power of the world to affect. A focus on affect does indeed highlight the centrality of the body and emotions but, most significantly, it not only incorporates both mind and body but also incorporates reason and passion by always presenting both sides of the causal relationship, so that mind and body, for example, are autonomous but correlate and operate in tandem. This opens up some new ways of looking at the relationship between mind and body which go beyond some of the phenomenological arguments and those based on body techniques addressed in Chapter 5, especially those based on the elision of mind and body, for example as expressed in Wacquant's work on boxing. A focus on affect requires consideration of the mind's power to think and the correlating body's power to act upon or in relation to those thoughts. This is a correspondence between mind and body rather than an elision and, as an open relationship, might also afford the possibility of disjunction, that is of a disruptive or uneven correspondence, because it is affects that cross the divide between mind and body and between actions and passions.

Sport also constitutes an example of affective labour in that sport serves to influence those for whom it is produced, a manifestation which is central to the sport media nexus with which this chapter is concerned.

Sports media: Making stories, making memories

The media play a key role in the creation of iconic moments, sports celebrities and major sporting events, and in everyday experience of sport. Sport is global not only because it is played across the world, but, more importantly, because the media transmit information across the globe so fast and so effectively to create a culture of sport and to place sport so prominently within popular culture. The media, like sport, are part of a massive global network that developed through the twentieth century and was strongly linked to the technological developments that made communications so important politically, socially and economically. Media messages are not simple, transparent reports of what is happening, on or off the field of play, however much many of them appear to be of numerical measurement. The parodic re-enactment of the hyperbole of sports commentary is informed by some evidential traces of the excitement that this commentary is trying to capture and recreate. The media are part of the whole process and experience of sport; they play a part in constructing our understanding of sport. Sport and its genealogies are made up of remembered feelings and emotions mediated by representational systems. Sport is itself constituted by a series of memories and conversations about victories and defeats, nearly winning and only just losing and pivotal moments. English football fans, bereft of success in the Men's Football World Cup competitions, have repeatedly rehearsed commentator Kenneth Wolstenholme's declaration: 'They think it's all over...it is now', after Geoff Hurst's third goal in the dying seconds of extra time in England's win over Germany in 1966. People not even born, let alone at Wembley stadium or watching the game on television recognize the phrase. It is the media's transmission of the moment that iteratively embeds it in the collective sporting consciousness and makes people feel as if they were there.

Media representations of sport and sporting events are closely tied to available technologies as well as the policies and practices of broadcasters. At the start of the twenty-first century it is television and the Internet that appear to have had the greatest and most transforming impact on sport with multinational and transnational companies such as BSkyB, Setanta and third-generation mobile phones extending media activities and increasing their influence. However, the influence of the mass media began with the printed word and print journalism retains an important place, especially in sports reporting and commentary (Mason, 1993).

The sports magazine industry in Britain was seen as developing from the end of the eighteenth century. Pierce Egan, who wrote about boxing and its associated activities, notably those involving various forms of gambling and other leisure pursuits of aristocratic gentlemen and their followers, was heralded as the first real sports journalist (Egan, 1812). Egan described the exploits of the *Fancy*, the term used to describe the amusements of sporting men from diverse social backgrounds, in the Regency period in the early nineteenth century in Britain, especially in London (Egan, 1812). Pierce Egan's pugilistic writing in the *Boxiana* series (1812–1829) described not only the theatricality of pugilism and the spectacle of fights, but also the social networks where identifications with masculinity through sporting, eating, drinking and gambling pursuits at times transcended differences of social class, but was dominated by the interests of those rich enough to enjoy its pleasures. Such writing highlights the sensual and sensate pleasures and dangers of sport, demonstrating that such meanings can be recreated through a synthesis of experience and textual representation even before the dominance of the sensational media practices and technologies of the later twentieth century. Its focus is also resonant of the networks of complicit masculinity that still dominate the media sport nexus (Woodward, 2007a).

The growth in sports journalism in the nineteenth century, with publications like *This Sporting Life* (1859), the *Sportsman* (1865), the *Sporting Chronicle* (1871) and *Athletic News* (1875), was facilitated by the steam press, the electric telegraph and the end of state taxes on newspapers (Mason, 1988), which meant that news could be collected and vast numbers of copies of papers distributed by train. Similarly, in the USA magazines on sport appeared from the 1820s, focusing on the sports favoured by the upper rather than the lower social classes. However, by the mid-nineteenth century, baseball began to acquire national prominence, with Henry Chadwick appointed by the *New York Herald* as a full-time reporter in 1862 (Smart, 2005). In both the UK and the USA, sport began to infiltrate spaces in the mainstream press too.

Sports writing continued to reflect the social experience of sport and its class-based affiliations. Tony Mason, writing about the sports media relationship in Britain, notes that sport began to take on its modern shape once it became part of the curriculum in public, that is independent, fee-paying schools (Mason, 1988). The sports press, in its early days, focused upon measurement, as a means of providing information about sports, giving details of venues, dates and times, and of educating the mass of the male population in the rules and regulations that governed sport as part of a process of education too by 'sporting evangelists' (Mason, 1988: 46). This 'evangelism' was part of the proselytizing movement to promote sport by an enthusiastic upper class among an already sporting working class, but with the additional emphasis on the advantages of the healthy body as linked to the healthy mind and hence the good citizen, as explored in Chapter 5. These regulatory practices also highlight the other aspects of Juvenal's original *mens sana in corpore sano*, the need to control excess. Embodied sporting practices may seem governed by regulation and discipline but sport is always haunted by the danger of excess and implicated in a diversity of affects. The healthy embodied self exercises moderation. Sport, given its corporeality, might offer pleasures that constitute excess and thus always carries a subtext of control that is also translated as moderation. Although early sports writers tended to be 'well-educated gentlemen, their prose

distributed in long columns, liberally punctuated by classical allu-
sions and quotations' (Mason, 1988: 46), sport has always brought
together different sections of society, and it has been an important
part of popular culture. Even in the early days of sports journalism, the relationship between the media and sport was not a simple
one involving the straightforward reporting of events, but strongly
invoked the pleasures of sport and its particular corporeal attractions.

Radio has a particular place in the sports media. Radio along with
cinema newsreels added greatly to the coverage of sport from the
1920s. The newsreels might have been spectacular, but radio was
more immediate. Listening to sport on the radio while it was actually happening created more excitement and the feeling of being
there at the ground. Thus radio sport gained the status of a 'fundamental social institution' (McChessney, 1989: 59) and it retains
a prominent position, for example through the advent of dedicated
sports channels such as TalkSport in the UK.

Radio's vital role in sports broadcasting is well illustrated by the
growth of the BBC in the early twentieth century. The BBC still
provides extensive, high-quality sports coverage in spite of the restrictions imposed on it by the growth of transnational commercial media
providers. Some of the most influential and powerful earliest media
sporting occasions were those presented on BBC radio, which started
in the 1920s. The development of sports broadcasting was closely
linked to technological developments and, on the BBC, sport developed along with outside broadcasting. The use of sport in promoting
the nation and, at the time, the British Empire was quite explicit. Key
moments in the sporting calendar were directly aimed at promoting
a sense of belonging to the British nation.

> Sport, of course, developed its own calendar very quickly. The
> winter season had its weekly observances of football, rugby and
> steeple-chasing, climaxing in the Boat Race, the Grand National
> and the Cup Final. Summer brought in cricket and flat racing, the
> test matches, Derby Day, Royal Ascot and Wimbledon.
>
> (Scannell and Cardiff, 1991: 279)

This is not just about technologies. The BBC and the BBC World
Service set up in 1932 permitted an acknowledgement and appropriation of the affective role of sport in quite direct ways that are

linked to belonging and becoming. Big sporting events could bring people together in what Benedict Anderson has called an 'imagined community' of the nation (Anderson, 1983). In the early days of the BBC and the Empire Service, which later became the BBC World Service, sports radio broadcasting provided not only a way of keeping ex-pats informed, but also, and more importantly, a way of defining Britishness and of recruiting citizen subjects by creating affects and affecting listeners. Broadcasting was integrated into the rhythms of the year and of the embodied listener.

> Nothing so well illustrates the noiseless manner in which the BBC became perhaps *the* central agent of national culture as its cylical role; the cyclical production year in year out, of an orderly, regular progression of festivities, rituals and celebrations – major and minor, civic and sacred – that mark the unfolding of the broadcast year.
>
> (Scannell and Cardiff, 1991: 278)

The BBC is 'noiseless' because the time-table is assumed and taken for granted as not only what *is* but what *should* be. Sport assumes the mantle of legitimacy which might also be sanctioned, by association with religion and the state.

This was framed by a discourse of fair play and impartiality but the merging of sporting events as part of the ritual symbolic of the remaking of the nation does suggest a synergy between rationality and passion in the process of becoming. Although this is the history of the service and the BBC retains its mission of transmitting high-quality, impartial material, the power of global sport has transformed its narrower associations with national identities, but although the context and power geometry have changed, for example, in cricket, the global focus of the traditionally English imperial game has shifted to India, Pakistan and Sri Lanka, through the massive growth of support and excellence in the Indian subcontinent as well as under the influence of powerful commercial interests. Sport has its own dynamic too.

Although radio continues to be an important medium, it is television that has been pivotal in the creation of sport as a transnational phenomenon. Television put sport at the centre of global culture and 'transformed the culture of sport, how players, coaches, owners and fans conduct themselves, the financial infrastructure of sport as well as the ways in which and the times at which games are played' (Smart, 2005: 84). In the UK, the BBC lost its monopoly when ITV arrived in 1955, although the BBC kept the big British events, such as the Boat Race, Test matches, Wimbledon, the Cup Final, the Grand National and the Derby, demonstrating an understanding of the affective power of sport broadcasting and the intersection of its rationalities in promoting national belonging and its passions through the appeal of sport's embodied practices. Radio is dependent on sound and the power of words to invoke the drams of corporeality as well, of course, as the visualization capacities of its listeners.These were the same events that terrestrial television endeavoured to hold on to when satellite and digital broadcasting came onto the scene.

Television provided a medium that could deliver the immediacy and visual drama of sport. This has had an impact of the sports themselves, whereby body actions are transformed through the discursive, symbolic field of representation. For example, when TV came to the USA, boxing and wrestling were broadcast early on. This was not only because these were initially technically the easiest sports to cover. Boxing has never lacked drama, but television created popularity for 'sluggers' rather than what had been seen as more graceful practitioners of the Sweet Science (Rader, 1984). Men's boxing was a very popular sport in the 1950s and it attracted sponsorship from a variety of commercial enterprises, notably breweries and tobacco companies. However, sponsors wanted to be associated with winners and not losers so the television coverage led to some mismatches (Rader, 1984). What might be construed as a poor performance on television could severely damage a boxer's career. An editorial in the boxing magazine *Ring* in 1955 suggested that television's greed for boxing had 'depleted the ranks of professional fighters by 50%' (cited in Rader, 1984: 46).

Television has, of course, advanced massively since the 1950s with colour, video technologies, communication satellites, action replay and slow motion, camera techniques and interactive facilities, along with the sponsorship that has accompanied and enabled such technologies. Just as national television made it possible to develop allegiances to teams and sports outside one's local area, global satellite coverage created the international fan. For example, satellite and digital television permitted the broadcasting of the English Premiership, Italian Serie A and Spanish La Liga all around the world, so that fandom has become less local, but offers new possibilities for collective viewing and new versions of identification. This is apparent, not only in the ubiquity of English Premiership strip on fans across the globe, but also because people gather to watch the match by 'uniting in front of their own television' (Leifer, 1995: 134).

The meanings of sport: Narratives and heroes

The print media, radio and television have enlarged the audience for sports and greatly enriched the experience of following sport. Sport provides the media with copy and content along with readers, viewers and customers. Sport provides many pleasures and creates and fulfils many desires, not the least through its creation of heroes, tricksters, villains, celebrities and fools, all of whom are constituted through the discourses of the media. Phenomenological accounts offer explanations in terms of the lived experiences of those who participate in the field of sport and focus more upon the routine practices than the iconography of celebrity. However, there is a sense of narrative structure in such approaches which resonates with media accounts and indeed with common sense. Media coverage of sport reproduces stories that are translated into compelling content for the news stories, magazines, television shows and Web sites that make up the media of contemporary popular culture. There are, of course, cultural variations, although there is a gendered weighting in the creation of heroic sporting figures but the media are central to their making.

Michael Mandelbaum (2004) uses the example of the US devotion to its three favourite sports to express a dominant view in the sports literature and among aficionados of mainstream sport about the role of narrative in the media sport nexus. He links sports to different stages of the US nation's development: baseball with the agrarian past, football as the model of the industrial age and basketball as post-industrial. Whilst Mandelbaum's argument may appear exaggerated in its oversimplifications, it does express a popular view of the power of narrative and 'the story' and an attempted synchrony between the material, embodied sports star or practitioner and the discursive field. This longing for clarity and coherence is in stark contrast to the complexities and multiplicities of postmodernist theorizing, but it does voice some of the perceived understandings of the appeal of sport, especially as temporally and spatially located, in this case in the USA. The creation of heroes is facilitated by what Mandelbaum argues is the adoption of epic form by baseball, football and basketball in which heroic protagonists overcome a series of challenges in order to accomplish their ultimate goal. He draws upon Homeric notions of heroism where great deeds are nothing if not recorded and whilst in ancient Greece storytellers created the narratives in the oral tradition, in the contemporary world it is the press which creates the heroic, or at least celebrity narratives. Second, he makes a play for modernity against the insecurities and multiplicities of postmodernity and suggests that these sports are attractive because they are coherent; they make life intelligible and comprehensible and provide a haven of security amidst the confusion of the post-industrial world. Each sport

> is a model of coherence for two reasons. Each is transparent: spectators can see for themselves what is happening and why. And each is definitive. At the end of each game, the spectators and the participants know which side has won. While the news section of the daily newspaper may report baffling and the unintelligible, the sports section features succinct histories that everyone can understand, with a clear-cut beginning, middle and end.
>
> (Mandelbaum, 2004: 8)

Mandelbaum argues that contemporary sport plays a significant role in modern societies by providing coherent stories, with 'shining examples' (2004: 10) of stars who are celebrities, heroes and role models in an activity that is much more than entertainment, because sports women and men really do what the spectators see them doing, thus invoking some reality principle in terms of the lived experience of sports fans:

> Interest in movie and sports stars goes beyond their performances on the screen and in the arena. Newspaper columns, specialized magazines, television programs, and Web sites record the personal lives of celebrated Hollywood actors, sometimes accurately. The doings of skilled baseball, football, and basketball players out of uniform similarly attract public attention. Both industries actively promote such attention, which expands audiences

and thus increases revenues. But a fundamental difference divides them: What sports stars do for a living is authentic in a way that what movie stars do is not.

(Mandelbaum, 2004: 10)

Mandelbaum's 'authentic' is a 'real' that encompasses the body practices of sport that belong in the lived experience of the body as situation. He quotes the baseball star Sandy Koufax as saying 'I don't think ballplayers are really entertainers... The customers come to hear the entertainers perform; they come to watch us live a part of our lives' (quoted in Mandelbaum, 2004: 11). The minutiae of the daily lives of sports stars are increasingly dissected and reproduced in the media (Andrews and Jackson, 2001) with considerable overlap between sports stars and the celebrities of popular culture. David Beckham, former England soccer captain, partner of a pop singer, fashion icon, simultaneously metrosexual and family man, role model and at the time of writing player at LA Galaxy, has crossed not only the Atlantic, but the boundaries of embodied sporting practices and sexualized physical display in popular culture.

Whatever his celebrity status and star rating in the terrain of popular culture, Beckham is more than an entertainer, because his sporting talent is the source of his heroic status. Sports stars often carry heroic status and are a source of 'admiration and emulation' (Mandelbaum, 2004: 11). There is an iterative process of identification which occupies diverse spaces through routine experience; in contemporary culture sports stars are also very ordinary super heroes. Other commentators distinguish between celebrity and heroism within the field of sport, demonstrating the difference between sporting celebrities and sporting heroes as part of a trend towards celebrity culture that is charged by the media.

Heroes and celebrities

Michael Jordan's retirement from basketball was greeted with some of the most expressive hyperbole in the testaments of the mass media.

God wore number 23 (*De Morgen*, Belgium) ...
God is going home (*Yedioth Ahrnonoth*, Israel) ...
The King is leaving (*Sport*, Spain) ...

King Mike abdicates (*Age*, Australia)
God will never fly Again (*Asahi Shimbun*, Japan)
God finally to retire (*Tochu Sports*, Jordan)
[His] name is engraved on the heart of everyone (*Beijing Morning Post*, China)
Ano Uno D. De J. (Year One After Jordan) (*Ole*, Argentina) quoted in 'The World Bids' 1999.

(Miller et al., 2001: 60)

These headlines confirm the celebrity and heroic status of a sports figure, who was frequently accorded deific status. Sport has often been compared to religion, especially in terms of the allegiances it enlists, so it is hardly surprising that its stars take on the mantle of the sacred.

The terms hero and celebrity have increasingly become used interchangeably, but they are fundamentally different. According to Daniel Boorstin,

> *The celebrity is a person who is known for his well-knownness*...The hero was distinguished by his achievement; the celebrity by his image or trademark. The hero created himself; the celebrity is created by the media. The hero is a big man [*sic*]; the celebrity is a big name.

(1992: 57, 61)

Thus, there are some clear distinctions between the two concepts and the challenge is to ascertain how and why they have become conflated.

While there are many cultural arenas in which individuals have emerged as heroes, sport has always been one of the key sites. Sport, as a cultural practice and institution, offers the opportunity for the demonstration of physical superiority in a system with clear rankings and rewards, the display of courage, commitment and sacrifice, and the chance to represent a particular group, community or nation. Sport heroes, both historical and contemporary, emerge from a wide range of personal achievements, social backgrounds and cultural contexts. A person might perform at a much higher level and with much greater consistency than the average; Sir Donald Bradman, Babe Didrikson, Jessie Owens, Paavo Nurmi, Pele, Nadia Comaneci,

Michael Jordan, Wayne Gretzky, Carl Lewis, Tiger Woods and Lance Armstrong (Ingham et al., 1993).

Mike Marqusee, however, suggests that athletic achievement in itself is insufficient to accord heroic status, citing basketball player Michael Jordan and the boxer Muhammad Ali to make this point. Jordan is famous, first, for making more money out of sports than any other athlete in history and, second, for his association with the Nike Corporation:

> The America, of which he is a symbol is corporate America and its winner takes all ethic. His blackness has been deliberately submerged within his Americanness, which is reduced, in the end to his individual wealth and success... There is... no way we can emulate Michael Jordan... In contrast we can all emulate at least some of what Ali did outside the ring... the adherence of conscience in defiance of social pressure, the expression of self through a commitment to a higher cause and a wider community. It was the willingness of the Greatest to link his destiny with the least and the littlest that won him the devotion of so many.
>
> (Marqusee, 2005: 295–297)

This also demonstrates another dimension of the hero versus the celebrity through the links between the body practices of sport and their social and political context. A particular performance might award heroic status, for example, Jackie Robinson, who, facing enormous racial discrimination and other social barriers in 1947, became the first 'black' athlete to play MLB.

Heroic status might be achieved by record breaking; Sir Roger Bannister who, in 1954, was the first person to break the 4-minute mile; or Sir Edmund Hillary and Tenzing Norgay (who did not receive the public recognition of the knighted Hillary) were the first to climb Mount Everest, the highest point on Earth, in 1953.

Some become heroes through risk-taking, personal sacrifice or saving a life. There are instances of athletes who have battled against cancer, like UK runner Jane Tomlinson or, more famously, Canadian Terry Fox, who was diagnosed with cancer, and with part of his right leg amputated, Fox set out to run across Canada in what he called the Marathon of Hope.

The heroes cited in the literature on sport are almost all male (Ingham et al., 1993). Increasingly, heroic status is something that is manufactured rather than achieved and heroism is becoming marginalized by celebrity (Rojek, 2001) with the media acting as arbiters of celebrity. The media are global, immediate and increasingly interconnected, resulting in a virtual saturation of celebrity culture linked to sport, music, fashion, film and reality television which blurs the boundaries, although this trend does not, as yet, seem to have provided more opportunities for celebrity status, let alone heroic, in sport for women. Male hegemony remains largely intact although the trend towards celebrity rather than hero is increasingly evident.

The intersections between sporting bodies and the media can be complex in their ethical implications as well as in the rather more crude complexity of the sport media commerce nexus. The BBC World Service combines an imperial past with impartiality, fair play and high-quality public service broadcasting and has to compete in a rapidly changing world where media privileges and access are only accorded through commercial advantage. Fast Track Africa provides a somewhat different manifestation of broadcasting and the example of an African footballer, who plays for his country and for an English Premiership team presents some of the complexities of representing heroes and celebrities and the different inflections of the construction of this figure.

Heroes and celebrities: Romantic visions?

I am a true son of Africa which is why I felt it important to take the BBC award back to the continent so I can share my success with them.

At the same time I hope to help young people receive the necessary attention, guidance and assistance that will empower and help them to fulfil their potential as productive citizens of Africa.

(Adebayor, 2008)

These are the words of the Togo and Arsenal football striker Emmanuel Adebayor, BBC African Footballer of the Year 2007. The phenomenon of African footballers who come to Europe sending money home is part of what makes football so attractive to aspiring youth. The other side of this is, of course, the exploitation of boys who hope to be able to do so and are 'discovered' by exploitative confidence tricksters who have no connections with the Premier league club they purport to represent as talent scouts. However, Adebayor's is a story of success. The question is whether the media sport relationship is necessarily one that creates only celebrities and not heroes and the extent to which reconfigurations of embodied sports heroes are possible and how they might be made.

In 2007 Adebayor won the Fast Track Sport Star of the year as well as BBC African Footballer of the Year, one of the most prestigious prizes in African sport, awarded after a public vote. The Togo striker garnered almost 42 per cent of the vote, beating Didier Drogba of the Ivory Coast into second place. The 2007 shortlist of five players was drawn up by football experts from every country in Africa, based on players' individuality, skill, technical ability, teamwork, consistency and fair play, a concept that has a long history in the construction of Britishness on the BBC World Service and in sport, albeit at some moments more strongly associated with upper-class Englishmen. On the BBC World Service programme, Adebayor was at pains to focus on his role in promoting a more socially inclusive cultural citizenship:

> I think a lot of people know me just on the pitch. They don't know where I come from and they don't know how I began. I put in a lot of hard work to be where I am today, but I'll never forget what it was like when I was young. Life was very difficult, and I told myself that I only had one chance to survive and that was to be a footballer. When I was going to Europe, I remember what my mother told me at the airport; she said: 'Manu, you see where we're living, you must go to France and do something good because we need your help'. When I was growing up I had someone to help me, to give me something, and today I'm in a position to help others, so helping people is always a pleasure for me.
>
> (BBC Press release, 2008)

One of the specificities of sport is that sport retains gendered distinctions which have been abandoned in other fields of experience and in other social relations. The male sports star whether a hero or a celebrity is still male, thus conforming to a bounded sense of self within the coalescence of a particular version of masculinity (Connell, 2005). Heroes, like stories, follow a pattern. Such identities

are frequently the focus of the media coverage of sport and presented in the form of a narrative. Donna Haraway argues, in pointing
to the limitations of the concept of narrative, however permeable
the boundaries, man is always the culmination of stories (Haraway,
1996). The hegemonic masculinity that remains the dominant trope
in such stories is not in itself a reason to abandon the narrative.
An alternative is to deconstruct such narratives and to posit different ones (Irigaray, 1989). There are also contradictory identifications,
where through the same apparatuses and practices and through
the same assemblages, binaries, including those of gender, are both
retained and challenged and the distinction between celebrity and
hero is one entry point for transformation in sport.

The celebrity is out of reach, the hero does things that we
might do (Marqusee, 2005). Adebayor is 'mobbed wherever he goes'
as an African footballer and a premiership star but he is 'not
showy' (Fast Track, 2008). More contemporary celebrity figures in
the Foucauldian sense of an historically emergent persona (Foucault,
1981) might appear to bridge the gap between the celebrity and the
hero as a role model like David Beckham, whose achievements for
his country on the football field are actually quite modest but whose
global standing is massive as a positive 'good citizen' role model,
suggests a reworking of celebrity which is more cosmetic than sub-
stantial. As Marqusee argues, the hero, like Ali, has to represent the
possibility of political change (2005).

Fast Track demonstrates some of the flexibilities and contingen-
cies of the BBC World Service as well as its role as cultural broker of
twenty-first-century global cultural citizenship. A responsive rearticulation of citizenship is still possible through the mechanisms of
public service broadcasting which is part of the make-up of the
assemblage of sport. Of course, sport retains its binaries, especially
in relation to gender, and sport is constituted through assemblages
of deeply inequitable relationships, notably as reproduced by the
commercialization of global sport, but sport also carries enormous
possibilities for participation and thus for social inclusion, which is
what reformed cultural citizenship can be. The BBC World Service's
legacy of impartiality and fairness have been reworked in the twenty-
first century so that in spite of its problems in competition with
satellite and commercial broadcasting it retains particular strengths
that can accommodate some of the tensions between celebrities

and heroes, even within the highly competitive field of sport in the remaking of cultural citizenship that addresses inequalities and marginalization. Sport offers a particular space in which different axes of power intersect, providing opportunities as well as exclusions, and Fast Track demonstrates a contingent intersection within the context of BBC World Service traditions.

Conclusion

This chapter has brought together some of the multiplicities of sport and, in particular, intersection between sporting spectacles as sensation and the sensations of spectators and has presented some alternative ways of bringing together the discursive spectacle and the embodied spectator. By suggesting that sport is an assemblage in which materiality, embodied sensations and discursive representations intersect, it is possible to challenge the idea that representations and symbolic systems are separate and distinct from lived, embodied experience, thus avoiding debate about which aspects of an equation or opposition carry greater weight. The dominance of the media sport nexus and volubility of celebrity in contemporary culture, especially as a vehicle for the further commercialization of sport, is not the only element in this assemblage, nor is it the determining force.

The media have played a central role in the development of modern sport which has been recognized in the discourse of sensationalism that has become a dominant motif of the media coverage of sport and analysis of the sport media nexus. Technologies and journalistic decisions, in whichever media, have combined to create possibilities of sensation, in the sense of shock and in invoking and creating desires and passions in the target spectator. Media practices have themselves shaped and recreated sporting embodied practices, including the construction of heroic figures and celebrities who embody the affects generated by media systems. The issue is not just sensationalism or what Deleuze called the cliché of sensation as spontaneous, however. Sensation is what brings the discursive field of media representation into the spectator and recruits the spectator into the embodied experience of spectatorship.

Phenomenological accounts emphasize the practicalities of being in the world and the lived body of the spectator or the follower of sport as they do the sports practitioner. Spectators, like practitioners, are engaged in an embodied practice and are drawn into the physicality of spectatorship, especially though the most dramatic representations or those that are strongly configured around a narrative. Watching sport raises the heart rate and is experienced through all the senses. This is why Joyce Carol Oates thinks you need to be at a fight to experience its authenticity, through the smell, the sounds, the sight and even the taste. To be at the match is to live the experience through the senses, evoked not only through the relationship between spectators and performers and through identification with the team, but also as part of the physical activity that is shared with other fans; the true fan who attends the games is not the opposite of the inauthentic fan who follows on the Internet or on television. A more meaningful, explanation of identification in sport is the immersion of the lived body in the sensations of spectatorship which constitute the affects of sport. Phenomenology focuses on personal, lived accounts, including those of the researcher and upon the lived body in the world; it has less to say about the discursive systems through which sensations are reproduced and recreated. It also presents problems in dealing with change and the dilemma of going beyond what is. Psychoanalytic theories present a radical engagement with what cannot be symbolized, as well as what can, but they too have some difficulty in challenging existing versions of the real, partly because whatever its disruptive potential, the unconscious remains embedded in a universal set of laws. There are, of course, challenges, for example, that focus upon difference but the starting point remains patriarchal.

Some developments of Deleuzian conceptualizations of affect, especially those which come out of psychoanalytical theories, present a reworking of emotion and feeling that might also be open to transformations and new departures as well as providing a nuanced reading of sensation and sensational that can overcome some of the entrenched binaries in sport and analyses of the sport media nexus. Rather than attempting to deconstruct sensation through the operation of unconscious forces or mediated narratives, this approach suggests that sensation is sensed rather than known and affect is a force which produces bodies. Sensation is not located in an individual body or embodied self; affects constitute a process. Affect is felt by the embodied subject but is also part of the technologies that produce it. The spectacle becomes the sensation in the spectator. Media technologies and the sport media nexus are affects too. Thus the discursive field becomes the affective field too. This approach goes a long way towards challenging dualistic thinking but still remains more focused upon the systems that constitute the processes than the enfleshed bodies that are both affected and the site of affects. Another way of explaining the relationship between enfleshed bodies and reg- ulatory bodies is through the lens of technologies of transformation which challenge not only binary logic, but also the boundaries of the body and other materialities and present an additional perspective on sporting embodied practices with a focus upon the possibilities of change.

7
Beyond Bodies

This chapter examines the extent to which body projects and practices facilitate transformations and the degree to which, in sport, such interventions may or may not transgress and contravene regulatory practices. The interventions and advancement of technological and scientific progress in sport as an example of the wider terrain of contemporary social life raises questions about the limits of what is human and how human subjects are constituted in this climate of change. The promise of monsters (Braidotti, 1996) and of cyborgs (Haraway, 1991), although rarely addressed within the field of sport, do inform sporting practices which range from technologies of health and nutrition, the technoscience of sports science to pharmaceutical interventions, across a range of sports and at different levels. Drugs and new technologies offer the means of overcoming some of the damage that the sporting body inevitably receives, but the promise of enhanced performance means such interventions can be as dangerous as they are beneficial; performance-enhancing drugs also suggest taking unfair advantage over clean, law-abiding competitors. Developments of cyborg thinking offer one set of explanatory mechanisms which can be negotiated within a framework of ethics and provide a means of addressing the issues that are raised by the promise, and threat, of technoscience.

The context of such debates has been framed by the corrosion of the moral underpinnings of sport and its values such as the principles of fair play and honour, the development of elitism and the concomitant subversion and undermining of the conditions that support sporting communities (Walsh and Giulianotti, 2006). Much of the debate addressed in this chapter focuses upon the framework of values, especially in relation to the fairness or otherwise of sporting practices. This raises questions about the operation of power within this discursive field in order to understand the constitution of subjectivities within the transforming power relations of sport by moving beyond the discursive construction of sporting bodies (Markula and Pringle, 2006) to address the boundaries of these bodies to assess the implications of what kinds of transgression are being made. Bodies are central to these debates; beautiful, fit bodies, super human and super athletic bodies and broken bodies, even monstrous bodies.

This chapter includes the crossing of diverse boundaries, for example, as identified by Donna Haraway in her work on human and non-human, including becoming machine and becoming animal (2005, 2008) and by Rosi Braidotti in her exploration of a materialist theory of becoming, drawing upon Deleuzian conceptualizations (2002) introduced in Chapter 6 in the discussion of the affective turn. Although Haraway starts from a very different position from the Deleuzian, she also posits a dislocation of the centrality of the human and argues for multiplicities and misplaced identities rather than linear, essentialized views of nature and biology. Haraway makes a powerful case against a Romantic approach to nature, the natural and the relationship between humans and animals, which is based on the idea of the post-human and of bio-centred egalitarianism (Haraway, 1991), which might still be criticized as an overestimation of the benefits of the post-human in this relationship. However, my purpose here is to consider the possibilities of going beyond bodies and how this might work in the context of embodied sporting practices.

Sport invites transformations of the body as well as providing a site where the boundaries of the human body can be interrogated and challenged as well as possibly reinstated. Self-regulation and techniques of the self not only suggest the transgression of body boundaries but also demonstrate tensions between regulatory bodies and the material bodies they target. This chapter explores how sustainable the reiteration of fluidity can be in explaining embodiment and the relationships between different materialities, including economic, social, political and cultural dimensions. Boundaries can be set through specific assemblages through which the embodied subject is constituted. The notion of the embodied subject as an assemblage incorporating the variables of class, culture, sex, race, ethnicity and disability which can be transcended, yet are at the same time sources of constraint, is well illustrated in the case of the disabled athlete.

Blade-Runner

'Blade-Runner' is the name given to the white South African runner Oscar Pistorius. His story, as played out in the run-up to the Beijing Olympics and Paralympics in 2008, demonstrates some of the ambiguities of embodiment and the intersections of different power axes in the classification of embodied sporting practices in the making and remaking of 'disabled' and 'able-bodied' athletes and transgression of boundaries between the two. Pistorius's renaming draws on the title of the film *Blade Runner* (Ridley Scott, 1982) and its concerns with the relationship between the human and non-human 'replicants', which, or maybe more accurately, who, can be seen as more human than the humans (Sobchack, 1999 [1987]). This film has been subject to critiques in film studies and especially feminist film studies (Doane, 1990, 1999; Penley et al., 1991; Sobchack, 1987) because of its relevance to cyborg thinking, for example, as developed by Donna Haraway (1991) and the gendered implications of othering and of the outsider, in the figure of the replicant. Women are frequently identified as the other and feminist criticism explains the attraction of cyborg thinking as challenging this positioning by reconstructing the relationship. The cyborg which transcends the binary logic of human and non-human accommodates difference and, as Mary Ann Doane argues, to know in *Blade Runner* is to be able to detect difference, 'not sexual difference but the difference between human and replicant (the replicant here taking the place of the woman as marginal)' (1999: 117). Difference can be constructed in a range of ways and through diverse processes, as was argued in Chapter 4, but difference is often characterized by 'othering' and binary oppositions. In science fiction difference is oftener produced through fantasy and the trope of the alien, both literally as in the *Alien* series films and in the dualistic language of this version of the imaginary.

Pistorius's science fiction epithet, which invokes the excesses and fantasies of science fiction, derives from the fact that the 21-year-old sprinter is a double amputee who lost both legs below the knees when he was a baby and runs on shock-absorbing carbon fibre prosthetics, which were designed by the Icelandic company Ossur 'to store and release energy in order to mimic the reaction of the anatomical foot/ankle joint of able-bodied runners' (Hart, 2008: 15). In 2008 he won the right to be eligible to compete at the Olympic Games in Beijing. The Court of Arbitration for Sport ruled that Pistorius be allowed to compete against able-bodied athletes. He had competed in two able-bodied athletics meetings in 2007, but the International Association of Athletics Federations (IAAF) ruled in January 2008 that his prosthetics qualified as technical aids, which were banned in IAAF-governed sports because they were seen to afford an unfair advantage to the athlete. In the end, he failed to qualify for the Olympics but ran in the Paralympics.

The Pistorius case, as first presented in 2008, was an unusual reversal of the more familiar media discourse of disability which constructs disabled athletes as vulnerable victims, albeit 'victims' who are courageous in defying the physical disadvantages they have experienced to compete in sport in the public arena. The language through which the debate was first constituted also invokes science fiction. Pistorius was renamed Blade Runner and the promise of cyborgification underpinned the initial more positive take on Pistorius's own achievements and the hope that his experience might offer other athletes with disabilities. Technology and ethics elided in a debate about the use of prosthetic blades described as the 'Cheetah flex-foot'; not only fast but it sounded like cheating. It was technoscience that the IAAF deployed to assess the limbs with a team of scientists using high-speed cameras, special equipment to measure ground-reaction forces, and a three-dimensional scanner to record body mass, prior to the IAAF decision to exclude Pistorius in Lausanne in 2007. On the one hand the interventions of technology might afford unfair advantage, but on the other, science provides the measure by which fair standards are judged and maintained. Ethical concerns and discourses of equity and 'fair play' intersect with the language and practice of technoscience. There is an uneasy dialogue between that which technology makes possible and an evaluation of who might benefit and who ought to benefit. Before Pistorius failed to qualify for the Olympics, Don Riddell of CNN (Cable News Network) described the 2008 ruling in Pistorius's favour as 'groundbreaking but his success might devalue the Paralympics bodied Olympics, asking 'Does it cheapen the Paralympic Games?' (Riddell, 2008). Some of those commenting of the BBC Sport 606 Web site were more realistic, asking if spikes will be banned next (BBC 606, 2008). Such commentators grounded the debate in an acknowledgement that sporting practices depend heavily and inevitably upon some form of technological intervention, even if only in the form of spiked running shoes. What is also important about Pistorius's experience as a competitive athlete is the challenge it offers to the parameters of the natural body and to what might be legitimate means of increasing body competences and achievements in sport and how and who judges what we can do and what we cannot.

When Blade Runner did eventually compete, in the Paralympics in 2008, there appeared to be even more emphasis upon the advantages that might be unfairly gained from technological interventions. Whatever the benefits, just or unjust, that might ensue from accessing the highest specification techno-aids, they cannot be available to everyone. The moral debate is located within the parameters of the economic and the social. Whilst the Paralympics are based on values of courage, determination, inspiration and equality, it could indeed be the last value, that is the most problematic, as is illustrated in competitions where some athletes clearly benefit from more advanced technologies than others. For example, in the 100 metres on 8 September 2008 there was a stark contrast between the rudimentary blades of Vanna Kim, a 40-year-old Cambodian who had lost his legs in a landmine in 1989, and Oscar Pistorius running on his state of the art Cheetah blades.

The Pistorius example raises questions about the theoretical positions which have emerged through such technologies. First, the promise of technology is at least partly subordinate to other materialities, notably the social and the economic. In the intersection of power axes, there remains a hierarchical weighting which challenges the fluidity of the systems within this assemblage and reaffirm elements of a more determinist paradigm. Second, technologies offer possibilities that transcend an essentialist notion of embodiment but the body is inextricably implicated in the constitutive aspects of subjectivity and retains some of the limitations of corporeality; what de Beauvoir describes as the ways the body as situation limits as well as enables our projects ([1949] 1989). Without underplaying the massive advances of technology and science and their liberating potential for bodies that are impaired, damaged, frail or sick, it might be possible to overstate the transcendental liberatory potential and the breaking down of boundaries between human and non-human; a point which raises the problem of essentialism if suggesting there might be a limit to the possibilities of transformation. Essentialism is, however, only a problem if it is construed as the opponent of progress and change and rejects all avenues of reconstruction and complexity. Recognition of material corporeality makes some transformations possible in ways that an insistence on the body as primarily the site of cultural inscription denies. Third, arising from the ghost of essentialism, the Blade Runner example poses a question about the classification of difference and how it is constituted, if, indeed, the concept still has any purchase if it is interpreted through the binary logic of normal/deviance, insider/outsider or nature/culture.

Techno-bodies

To what extent does technoscience de-materialize the body and render the enfleshed body with all its limitations redundant? New technologies clearly have enormous potential for enhancement. Whatever the level of intervention the cyborg, even in representational forms, as in cinema, retains its human dimensions. Arnold Schwarzenegger in the Terminator series, however, constituted by metal and wiring still has the body of an extremely muscular fit body-builder. 'The cyborg is a culturally dominant icon whose effects go well beyond those of cinematic and media seduction. They affect corporeal behaviour of "real" humans the world over...with their silicon implants, plastic surgery operations and athletic-like training' (Braidotti, 2002: 244). The stars of popular culture like Madonna may adopt athletic-like training to sustain their bodies and conventionally sexualized visible appearance, even into their fifties. In sport athletes are not 'athlete like', they are actual, lived bodies or bodies as situations in de Beauvoir's sense (1989). However, this is not a simple juxtaposition of the semblance of athleticism and the actual competence of the embodied practices of athletes. The embodied practices which reproduce the athletic body are the material incorporation of the body as situation with athletic skills which may be classified through the regulatory bodies of sport or through a sexualized discourse of representational norms and practices. The body practices and embodied selves are not constituted within an opposition between masquerade and actual, between representation of what could be a semblance and real sport, but through the intersection of different axes of power which embrace different materialities.

However, cyborgs may posit a post-human, that is, artificially reconstructed body (Balsamo, 1996) so that the body no longer fits the naturalistic paradigm, thus creating the Foucauldian paradox of simultaneously being everywhere and very visible and also disappearing in the shift from flesh and blood into technological constructions. Blade Runner would appear to conform to some of this configuration.

There are problems with the overemphasis on the cyborgification of bodies which are highlighted by a consideration of cyborg bodies in sport. Sporting bodies are indeed everywhere in the discourses of popular culture, where they play different parts, as super-fit, super-human heroes and as vulnerable victims of a culture of hyperbole. The so-called post-human embodiment is also part of the cash-nexus and of media, sport, commerce networks. Bodies are everywhere, but the bodies we see are not those of real people, whose bodies are marked by the boundaries of the self as natural entities. 'Real' people do, of course, pose a dilemma if 'real' is constituted as in opposition to unreal or the illusion of representation as was discussed in Chapter 6; they could also be *irreal* (Sobchack, 2004). Deleuze avoids this danger by challenging the very idea of signification and representation, which has been so central to the semiotic critiques of cultural studies. Representations are, of course, crucial to the meanings that are attached to sport, but Deleuze suggests an alternative to semiotics by bridging the divide between the representation and what it is claimed to represent by merging the two. For example, Deleuze suggests that all becomings are real even through being an actualization of the virtual. The virtual is opposed not to the real but to the actual:

> 'real without being actual, ideal without being abstract'; and symbolic without being fictional. Indeed the virtual must be defined as strictly a part of the real object – as though the object had one part of itself in the virtual into which it plunged as though into an objective dimension.
>
> (Deleuze, 1994: 208–209)

If this means that the symbolic is in the embodied, it could create an assemblage that incorporates enfleshed and discursive bodies, but Deleuze's disavowal of the actual makes this unlikely within his philosophical framework.

Technologies are strongly associated with change and especially the speed of change in contemporary society, especially in global sport, and much of the theoretical developments on both embodiment and identities and selfhood have sought to challenge notions of fixity and essentialism in espousing the liberating possibilities of change. What have somewhat loosely been called post-structuralist critiques start with either a fragmentation of the unified subject or of the idea of multiplicities which challenge the view that the embodied self is a given or has essential qualities; with unity goes the notion of an essence, both of which are rejected within postmodernist, post-structuralist approaches. Some feminists have taken up a Deleuzian approach in order to how transformations occur and to provide a radically new theoretical framework that can accommodate the speed of technological and scientific change. Braidotti argues for the benefits of Deleuzian approaches which rethink the human subject in a non-essentialist but vitalist philosophy which deploys the idea of nomadology as a philosophy of immanence which rests on the notion of sustainability as a principle of containment and development of a subject's resources, understood environmentally, affectively and cognitively as systems within the assemblage. This philosophy purports to combine the specificities of the subject that is aligned to the mobility of becoming. Thus a non-unitary subject inhabits a time that is in the active sense of continuous becoming. Deleuze argues against linearity so that Becoming 'is a movement by which the line frees itself from the point and renders points indiscernible ... Becoming is anti-memory' (Deleuze and Guttari, 1987: 291).

This philosophy is anti-hierarchical and suggests that becoming is a mapping and a connection rather than a linear hierarchy; thus difference too is not oppositional or weighted. Difference is construed as a mutual becoming. However, it is worth noting that 'becoming' applies to women and animals. Deleuze and Guttari present the notion of 'becoming woman' because they take man as 'the ground of becoming' (1987: 291), which clearly presents problems for feminists becoming Deleuzian. Braidotti, however, asserts that what Deleuze formulates is a philosophy rather than a mode of description. Braidotti argues for the use of Luce Irigaray's theory and politics of difference as a means of using the radical rethinking of change and becoming in Deleuze, but retaining the understanding of difference that belongs in Irigaray's work (1991).

The question for theorizing techno-bodies is how the notion of becoming could be applied to the transformations of embodied selves in the field of sport and what this concept can deliver in explaining how the boundaries of the body can or cannot be subverted and transcended.

Deleuzians stress the primacy of change and of the crises which characterize the contemporary world with human becoming at the centre of technological and scientific advances that question what makes up life itself. As Brian Massumi says, 'the "human" is more closely akin to a saleable virus, neither dead nor alive, than a reasonable animal standing at the pinnacle of earthly life-forms, one step below the divine on a ladder of perfection' (Massumi, 1998: 60). The seemingly post-human disorder of such a claim of constant flux is both alarming and exaggerated and there is some degree of hyperbole in such claims that so overemphasize the extent of constant change. However, what does appear to be a major challenge in the contemporary world is the centrality of technoscientific interventions, which provide opportunities for creativity and invention as well as subversions and constraints. Some of the complex interfaces of technologies that have life as their focus do challenge traditional ideas of being human. As Anne Fausto-Sterling argues, 'we live in a geno-centric world' (Fausto-Sterling, 2000: 235).

Sport is part of this world and a space in which the techno-hype of regulating bodies has an uneasy presence. On the one hand the developments of technoscience provide immense opportunities for creating ever faster records and increasing the stamina and strength of athletes. On the other hand such interventions create regulatory dilemmas, which are haunted by nineteenth-century notions of fair play and the threat of monstrosity. Pharmaceutical interventions are part of the discourse of fair play and present a particular problem in relation to the speed of change. The failure and inability of sports bodies to grasp these changes is understandable; their criteria are out of date before they have time to formalize them, let alone implement them. New drugs mask the effects and manifestations of previous performance-enhancing substances and this is a confused and confusing terrain.

Sports bodies are made through corporeal practices and subject to the frailties of human flesh, even if those at the top of their professions can depend on the best treatment and most highly sophisticated interventions of medical science. The regulatory bodies of sport grapple with the ever-developing technologies and pharmaceutical possibilities, endeavouring to establish boundaries in a field of mobile uncertainties. However, there are instances when the body fights back and reasserts its enfleshed materiality. This can be through the actual achievement of embodied selves within the parameters of governance where body practices re-instate themselves. The US swimming star, Michael Phelps, who won gold medals in five events at Beijing in 2008, achieved success through rigorous routine practices. His training and nutrition regime and certainly his ensuing success may be denied to most other people, but he is not a mythical construction, or a fantasy, and this body is reproduced through the routine, iterative embodied practice of a dedicated training (and eating) regime (Phelps, 2008). Bodies also fail for human, enfleshed reason, for example, through sports injuries. The conceptual fight back sometimes takes the form of corporeal defeat which results in the broken body.

Broken bodies

Sport highlights the inadequacies of reading the body as a text and a site of inscription and presents multiple points at which discursive inscriptions are challenged whether this is understood from Foucauldian or Deleuzian perspectives. The athlete's body demands acknowledgement of physical materiality, which questions the claim that the body is a surface where forces play and 'there is nothing essentialist or primordial about Deleuzian bodies' (Braidotti, 1991: 74). The exchange in sport often takes place within the context of the relationship between the beautiful body that incorporates not only physical fitness, muscle, skill and good looks, but also a whole set of physical experiences that constitute the energies of the sporting self and the damaged body with all its points of frailty. The athlete's body includes both materialities.

Boxing presents its own versions of these materialities. Damage ranges from the more personal experiences of minor injuries to death, such as that of Bradley Stone and brain damage as in the case of

Michael Watson. The year 2005 saw the first death of a woman in the ring when Becky Zerlentes died at the age of 34, after being knocked out in an amateur bout in Washington. As the artist Sandor Szenassy (1995), whose work includes boxers as subjects and whose patron was the boxing promoter, Frank Warren, is quoted as saying, 'People think we live in a civilized society. Boxing hurts our sensibilities; it reminds us that we don't'. Boxing hurts our sensibilities because of its physical materiality and public display of hurt bodies.

Boxing abounds with events which challenge civilized sensibilities. Oliver McCall suffered what was described as a nervous breakdown in the ring at his WBC heavyweight fight against Lennox Lewis in Las Vegas in 1997. 'At the end of the fourth round, when McCall threw just two punches, he was crying . . . and needed just 55 seconds of the fifth to see that the man's problems were deep-rooted and mental' (Mossop, 1997: S7). The anatomical body is inextricably central to the experience and the representation of boxing. McCall's body refused to respond from round four. This is conceptualized as 'mental failure' but what was evident is that McCall could not fight. This example illustrates the passive rather than the active synthesis of mind and body which is visible in the broken body in the ring in an unusual scenario. McCall became an object of pity and not of desire or even hostility in what was perceived to be a loss of control of his body.

In the fight between Nigel Benn and Gerald McClellan in February 1995, in a bout described as Benn's greatest and his worst night (Mitchell, 2003), it was McClellan who suffered. After the fight McClellan remained in an induced coma for 11 days while doctors eased the blood clot on his brain. McClellan no longer knows about boxing titles or records and can no longer see (Mitchell, 2003). Much of the debate has been framed within a moral dilemma about whether the film of the actual fight should be shown, for example on television after the event, but it also focuses on the materiality of the flesh and the broken body which is the experience of the protagonists whether or not it is subject to representation or interpretation.

Broken bodies are nonetheless constructed through the prism of beautiful bodies in sport. The beautiful body draws upon an aesthetic that has a long history, especially in bringing together notions of physical beauty, fitness and rectitude in a discourse that has often been naturalized. The juxtaposition of beautiful and damaged bodies is embedded in the genealogy of sport none more so than boxing, especially in its configuration of masculinities. These are strongly gendered beautiful male bodies. Women's sporting bodies as part of a cultural aesthetic are more recent in the public arena and almost completely absent from the classical past. In contemporary culture women athletes' bodies are still more likely to be sexualized when represented outside the sporting arena (although, of course, men's bodies have now entered the realm of sexualization and eroticization, led by England, LA Galaxy and AC Milan footballer David Beckham).

The aestheticized male body was reproduced in relation to classical ideals following the eighteenth-century revivals of the ideals of Greek sculptures, which include the figure of the boxer. The two best-known statues are the bronze statues of the *Terme Boxer* and the *Belvedere Torso*. The *Terme Boxer*, a seated figure in a contemplative pose, is cited as masterpiece of Hellenistic athletic professionalism. His head is realistic with cropped hair, low forehead, broken nose, cauliflower ears, numerous facial scars and a mouth suggesting broken teeth (Smith, 1991: 54).

Body parts are the preserve of damaged bodies here too. The marble *Belvedere Torso*, on the other hand, is accorded nobility and grandeur from the realm of myth (ibid.). The point is made that the latter is a classical heroic figure whereas the *Terme Boxer* 'despite all his training and physique remains firmly earthbound' (ibid.: 55). Bodies are both discursive through fantasy and myth and material, 'earthbound', in fact, and the body is the route into finding out how they interrelate.

Muhammad Ali at his peak incorporated the beautiful body, but the ill health which has dogged his later years, albeit which he denies to have arisen from his boxing experience, presents another dimension of embodiment. A documentary film *True Stories: Thrilla in Manila* (John Dower, 2008), which presents Smokin' Joe Frazier's sad story of bitterness, portrayal, black politics and racism, rearticulates the violence of the particularly brutal heavyweight fight between Ali and Joe Frazier in Manila in 1975 as divine retribution. Although Ali won the fight and the title and heroic status, whereas Frazier, in spite of his own successes now lives modestly above his boxing gym in a run down part of Philadelphia, it is Ali whose body is most impaired through the damage of Parkinson's disease which Frazier attributes to the blows he inflicted to his opponent's vital organs on that day in the Philippines in 1975. Frazier describes the blows to Ali's liver and kidneys that are recorded in the film's coverage of the commentary as an event, a sensation in the Deleuzian sense rather than as part of the process of signification. Bodies have parts, which are material. However, damage to the flesh is not merely an event; it has material consequences that are both built into the narrative reconstruction and, more importantly, the body as situation.

This sporting event invoked monstrosity in its rhetoric even before it had taken place. In the lead-up to the fight Ali had repeatedly referred to Frazier as a gorilla. The rhetoric galvanized public interest and attention and also demonized Frazier, which demonstrates both the negativities of monstrosity and a fight back, which is expressed through the body. Boxing is characterized by binary logic but this is a non-linear narrative in an assemblage of embodied selfhood that reconstituted at different moments and through multiple processes. Ali is a heroic figure, as recreated through legends and rewritten stories and in a dialogue between the beautiful body and the tragic enfeebled victim. In this drama, however, Ali is remade as hostile and dishonourable and suffers physical impairment which is presented as a just punishment. Even in a sport like boxing there is a sense in which difference is reconstructed and always mobile. This is not a linear narrative, but there are stories and memories which are in play in the reconfiguration that is taking place and it is difficult not to address the centrality of a story, however often it is retold in this assemblage. This situation is striking because of its invocation of body parts to demonstrate.

Mike Tyson, like Ali, at his peak, appeared both invincible and physically perfect whatever his emotional shortcomings (O'Connor, 2002). Tyson presents a different assemblage. Beautiful boxing bodies might be said to comply with an ideal that equates physical beauty with health, success and strength, both moral and corporeal, but they are always subverted by the threat of damage and the broken body. Rather than opposing essentialism with a breakdown of all boundaries, sporting bodies offer a means of showing the intersections between different dimensions of the discursive and the enfleshed body and between points of contact and disruptions, through the tension and the interrelationship between beautiful and damaged bodies.

It is through the material body and its practices that heroic status is achieved, rather than body and myth occupying separate spaces. The power of narrative in the constitution of embodiment has to be incorporated into a theoretical framework through which to understand embodiment and the relationship between the discursive and the enfleshed body because it is through the stories that become myths and legends that difference is constituted. Sporting bodies are recreated through gendered narratives. *The Thrilla in Manila*, however retold and reconstructed, is about the constitution of masculinities and an unstated gender difference as well as a difference between broken and beautiful bodies. Frazier and Ali both have daughters who have fought but Laila Ali and Jacqui Frazier are neither mentioned in the film, nor accorded anything like the status of men, not because they have less powerful, smaller bodies than their fathers but because they are not part of the narratives of greatness. Laila Ali, for example, presents the magnificent body of a strong black woman, but apart from images set in the context of her boxing achievements, she is more likely to be represented through the lens of familial relationships; whether through her even more famous father or as herself a mother, thus reaffirming her status in relation to both heterosexuality and kinship ties. Even Olympic women athletes are expected to conform to heterosexualized criteria of glamour off the field. British swimmers and cyclists who achieved gold medals at Beijing were then photographed facing the challenge of killer heels. Athletes like swimming gold medallist Rebecca Adlington embody success and the incredibly demanding training regimen required to achieve so highly and the ordinariness of both a local girl who made good and a woman who inevitably wants to look sexy and wear stiletto heels (Adlington, 2008).

Material bodies matter but so do the stories that are told about them. Assemblages of embodied selves have to include the flesh and the body parts that constitute the self as well as the stories that are told about them. Such an assemblage is more than a set of systems or forces that inscribe bodies.

If it is only women (and animals) that are becoming because men, or rather man, is the starting point and is not in the process of becoming, there are different possibilities. This may be read positively as opening up the opportunities for women (Braidotti, 2002; Grosz, 2000) or more negatively as yet another reinstatement of what is and what has yet to be challenged. Deleuze does suggest that by losing his centred identity when, as he expresses it somewhat wittily, the subject man becomes the object, in a loss that could be energizing.

Either way there is an assumption about man as the grounding for what is that is clearly manifest in sport, but which, I suggest, can only be deconstructed and addressed by a focus upon difference which embraces the specificities of differentiated embodiment. Women risk being relegated rather than elevated to the corporeal. As women, they are seen as subject to the whims, vagaries and cycles of the body, rather than achieving the control and mastery (*sic*) associated with the mind and with masculinity. When masculinity is embedded in corporeality as in a sport like boxing, what is emphasized is the control and discipline of the body rather than any unregulated excess of physicality, however violent the disruptions that may lead to broken bodies may be. Cyborgs are also gendered and marked by difference. Cyborgs are largely white and male; a product of the masculine symbolic (Irigaray, 1984) even though they purport to blur the boundaries and even to eliminate such distinctions. The post-gender world has not yet arrived whatever blurring of boundaries and the optimism of cybernetics and Deleuzian becoming man-woman-child, although both offer a route into rethinking difference.

Gendered selves

The materiality of enfleshed selves is powerfully implicated in the construction of difference. Participants in sport are classified by their bodies and body measurement; almost all sports differentiate by gender. Whatever the fluidity of gender realignment in the wider spaces of Western societies and the liberal argument that people are the gender they perceive themselves to be, sport has its own rules and measures and seeks to impose what in dominant sporting discourses constitute standards of 'fair play'. Technologies of change pose dilemmas though for those who seek certainty and destabilize the boundaries of embodied gender as well as the merging of human and machine and biochemical make up. This is well illustrated in the history of 'gender verification' at the Olympics. Historically, gender has been classified by physical characteristics and more recently by DNA, the fallibility of which has only recently been recognized in sport, with the IOC stopping genetic testing only in 1992.

While there has been abundant publicity regarding the testing of Olympic athletes for use of prohibited performance-enhancing substances, for more than 30 years the IOC has required all female competitors to undergo gender verification. The purported rationale is to detect male imposters who would have an unfair competitive advantage.

Women athletes born with relatively rare genetic abnormalities that affect development of the gonads or the expression of secondary sexual characteristics have presented a problem for the measurement of gender. In part, the controversy over gender verification reflects the increasing popularity of women's sports. The ancient Olympic Games were limited to men, who competed in the nude. Women spectators were prohibited. De Coubertin was opposed to any women competing at all.

The IOC introduced sex testing in 1968 at the Olympic Games in Mexico City, after the masculine appearance of some competitors. Gender-determination tests were seen as degrading, with female competitors having to submit to humiliating and invasive physical examinations by a series of doctors. Later the IOC used genetic tests, based on chromosomes. Geneticists criticized the tests, saying that sex is not as simple as X and Y chromosomes and is not always simple to ascertain, because, for example, it is thought that around one in 1000 babies are born with an 'intersex' condition, the general term for people with chromosomal abnormalities. It may be physically obvious from birth – babies may have ambiguous reproductive organs, for instance – or it may remain unknown to people all their lives. At the Atlanta games in 1996, eight female athletes failed sex tests but were all cleared on appeal; seven were found to have an 'intersex' condition. As a result, by the time of the Sydney games in 2000, the IOC had abolished universal sex testing but, as will happen in Beijing, some women still had to prove they really were women.

Transsexuals, who have had a sex change from male to female, can compete in women's events in the Olympics, as long as they wait two years after the operation. Then humiliation of gender testing is evident in several of the high profile cases of recent years. For example, Santhi Soundarajan, a 27-year-old Indian athlete, was stripped of her silver medal for the 800 metres at the Asian games in 2006. Santhi, who has lived her entire life as a woman, failed a gender test, which usually includes examinations by a gynaecologist, endocrinologist, psychologist and a genetic expert. It is likely that she has androgen insensitivity syndrome, where a person has the physical characteristics of a woman but the genetic make-up includes a male chromosome; the trauma of the testing led her to attempt suicide while awaiting the results. Edinanci Silva, the Brazilian judo player, born with both male and female sex organs had surgery so that she could live and compete as a woman. According to the IOC, this made her eligible to participate in the games and she competed in Atlanta 1996, Sydney 2000 and Athens in 2004.

Other instances are more explicitly politically motivated, like Dora Ratjen, a man who was forced to compete at Hitler's 1936 Olympics. He ultimately lost the gold medal in the women's high jump, having been later found to be Herman Ratjen, a member of the Hitler Youth. These instances of gender ambiguity are framed within ethical discourses with the implication of cheating and moral transgression and usually involve either a reassertion of the masculinity to which women, mistakenly because they cannot attain it without being masculinized, aspire in sport a sense of pity. More recently, in the former East Germany it is estimated that as many as 10,000 athletes were caught up in the attempt to build a race of superhuman communist sports heroes using steroids and other performance-enhancing drugs. The shot-putter, Heidi Krieger, was given steroids and contraceptive pills from the age of 16 and she was an European champion by the age of 20. Her overdeveloped physique had put a huge amount of pressure on her frame, causing medical problems, while the drugs had caused mood swings, depression and resulted in at least one suicide attempt. Later Krieger underwent gender reassignment surgery claiming that she had been confused about her gender, but felt that the drugs had pushed her over the edge.

The US, Polish-born sprinter Stella Walsh who won gold and silver at the 1932 and 1936 Olympics, respectively, and set over 100 records, was found at her death to have male genitalia, although she had both male and female chromosomes, a genetic condition known as mosaicism.

In 1990 the International Association of Athletics Federations (IAAF) became the first major international sports body to recommend allowing transsexuals to compete, with some restrictions which were agreed in 2004. Athletes who have sex reassignment surgery before puberty are automatically accepted as their new sex must have all surgical changes completed, be legally recognized as their new sex in the country they represent, and have had hormone therapy for an extended period of time. For male-to-female transsexuals, this generally means a minimum of two years. These examples demonstrate some of the different aspects of rethinking the boundaries of the sexed body.

The debate is situated within a discourse of 'fair play' and the prevention of the possible advantages which male athletes might gain by participating in women's competitions, barely concealing patriarchal discourses which masculinize any successful female athletes and fail to recognize the possibility of women's athletic achievement; if they are any good they must be men. Cyborgification is weighted in sport in favour of a male norm and regulatory bodies are keen that participants should not gain unfair advantage, either through pharmaceutical interventions, or, having accumulated the upper body strength of a man who then seeks to compete at women's events either by passing, or in the very unlikely and rare case in sport of gender realignment. Although there is an increasing recognition of the limitations of genetic testing, chromosomes do not complete the picture and the emphasis has shifted onto pharmaceutical interventions to enhance performance.

Pharmaceutical sporting practices

Drugs present another dimension of cyborgification. Performance enhancement by pharmaceutical means is implicated with other means of improving performance whether through appliances or attempts at passing, usually by men as women. Regulatory bodies have concentrated on athletes' use of performance-enhancing substances for a variety of reasons, including the damage which athletes themselves can suffer, for example female gymnasts taking drugs to delay puberty or body builders taking anabolic steroids. All sports and all sports systems are involved so much so that the use of drugs has become a key area where regulatory bodies and the bodies that participate in sport are mutually constitutive. Sporting bodies disagree about the governance of drug use in sport but all are implicated. This is illustrated by the inclusion of myriad aspects of drug use at the WBA conference in 2008.

The 87th WBA Annual Convention in Punta Cana, Dominican Republic, closed with a medical seminar, summing up the issues discussed during the week, which included managing drugs in sport, the issue of why a subdural haematoma should not automatically disqualify a boxer from returning to the ring, the security and clinical factors to be considered of female boxing, metabolic alterations in bone and endocrinological alterations in women boxers, and muscle physiology in sport practice (WBA, 2008). There may be an element of paternalistic concern about women's use of drugs and these issues are strongly gendered with more emphasis on caring for female boxers and a stress on allowing even male boxers who have experienced subdural haematomas to continue to fight, but there is a strong focus on the dominance of pharmaceutical issues in sport overall.

One of the bodies with most concern has been the IOC. In 2008 in the run up to the Beijing Olympics concern was expressed about the increase in the number of drugs available to enhance performance and the problems the IOC was experiencing in detecting drug use. The debate was framed in the language of fair play. For example, an account of the previously undetectable drug use by cyclists, Cera: Continuous Erythropoiesis Receptor Activator, which has almost become part of the rules of the sport, as manifest in the *Tour de France* was described as follows:

> Like most, if not all performance-enhancing drugs, EPO Cera was intended for genuine therapeutic purposes. This drug was developed by the Swiss pharmaceutical giant, Roche, for the treatment of chronic kidney disease and is a form of EPO that is longer lasting and requires fewer doses... For cheats more used to frequent injections and short-term boosts, the advantage of using EPO Cera is clear, even though Roche have stated that its 'misuse by healthy people could lead to an excessive increase in haemoglobin which may be associated with life-threatening heart problems'.
>
> (Knight, 2008)

Drugs that improve natural potential are against the spirit of this model of sport, although what is natural is problematic, since athletes have always sought ways of improving performance, and boundaries of acceptability are not always clear or are too difficult to construct. Some training programmes imposed on vulnerable young competitors might also cross the boundaries of fair play. Biological bodies are largely assumed in sport though and are mostly configured around categories of difference, especially as demarcated by age and gender. Fair play is important and performance enhancement by means that are deemed by regulatory bodies as beyond the parameters of accept-able practice and might endanger athletes' well-being transgress these regulations and some athletes are more vulnerable than others to the pressures of competition, governance and the commercialization of sport. The discourse of care which surrounds this debate is also informed by a notion of agency and the idea that athletes might choose what kind of training to use and how to compete, although this presents a somewhat naïve view of the sports, sponsorship and media complex.

Sex/gender

Debates about difference have often centred upon gender, initially as a discussion of sex and gender, for example within feminist theory, especially in relation to the problems of bodies and in the con-text of the hierarchical nature of that relationship when the two are presented as separate concepts. Whilst the mind and the soul might be rated above the body in soul/body, mind/body dualisms, in the sex/gender debate, the embodied sex has greater weighting as a determinant of gender. Liz Stanley (1984) described the argu-ment as being one between biological essentialism, which prioritizes biological, embodied sex as the determinant of femininity or of mas-culinity, and *social* constructionism, which focuses on gender as a social, cultural category. Sex was associated with biology (and subject to testing by the IOC) and gender with social and cultural practices. Sex and gender have been combined, but there is still the assump-tion that sex as a biological classification is privileged over gender as covering social attributes, in terms of the certainty it affords in relation to identity. Second, where the two have been explicitly disentangled, the influence of sex upon gender has been awarded priority and higher status than any influence gender as a cultural and social construct might have over sex. There is also a normative claim involved in this hierarchy, namely that sex *should* determine gender.

Sex and gender have been elided to women's disadvantage, whereby cultural expectations of what was appropriate or possible for women were attributed to some biological law (Oakley, 1972). The notion that women should be relegated to second-class citizen-ship, or even accorded no citizenship status, because of anatomical difference from men, in particular, the possession of a uterus, has a long history in sport. The tension between acknowledgement of women's bodies as situations manifesting particularities that require different provision of care and the subordination and devaluing of women justified on such grounds has been confused and in some ways reflects the equality difference debate discussed in Chapter 4. Women have been excluded from activities ranging from sport (Hargreaves, 1994), because of their 'sex', which was claimed to be generative of dire outcomes such as Aristotle's 'wandering womb' and its contemporary translation into the protection of reproductive capacities. Women have been excluded from sports and continue to run shorter distances, play bouts of less duration and comply with different regimes from men in sport, such as playing off different tees in golf or fighting fewer rounds in boxing, on the basis of physical difference. Feminists sought to make a distinction between the biological characteristics of the body, the anatomical body and gender as a cultural construct. However, more recently, the idea of an oppositional distinction between sex and gender has been challenged, most powerfully by postmodernist feminists such as Judith Butler (1990, 1993) and for many the term 'gender' is largely preferred. Sex too is discursively constructed (Butler, 1993). The meaning of 'sex' is strongly mediated by cultural understandings that, it is argued, make it impossible to differentiate between sex and gender and it is increasingly recognized that they are mutually constitutive. The use of gender permits an acknowledgement of this powerful cultural and social mediation (see Proce and Shildrick, 1999).

Toril Moi suggests the 'lived body' an alternative to the categories of sex and gender (de Beauvoir, 1989; Moi, 1999). This concept provides a non-essentialist synthesis of the physical body and its experiences in the world around it and initiates action in different situations which can thus be differentiated. The body is located in a given environment so that, as a 'situation' incorporating the physical facts of its materiality, such as size, age, health, reproductive capacity, skin, hair and the social context, Moi's 'lived body' is not biologistic, that is, it is not reducible to its corporeal parts, subject only to general laws of physiology and divided into two categories of gender. This eliminates the constraints of other binaries too, such as nature/culture; the body is always a part of culture, inculcated with habits, acting according to social and cultural rules.

To consider the body as a situation...is to consider both the fact
of being a specific kind of body and the meaning that concrete
body has for the situated individual. This is not the equivalent of
either sex or gender. The same is true of 'lived experience' which
encompasses our experience of all kinds of situations (race, class,
nationality etc) and is a far more wide-ranging concept than the
highly psychologizing concept of gender identity.

(Moi, 1999: 81)

The idea of the 'lived body' provides one route into addressing difference without reductionism and for the exploration of the othering
that arises from the marginalization that accompanies difference.
However, as Iris Marion Young (2005b) points out, although the
lived body avoids the binary logic of sex/gender, it may pay insufficient attention to the structural constraints which shape experience.
Consequently, she argues for the retention of the concept of gender.
Embodied difference is not only a constituent of subjectivity and the
formation of the self and part of the explanatory framework through
which gendered identities can be understood, it is also part of the
gendered division of labour and requires the possibility of collective
action, for example, in challenging social exclusion and its embodied
consequences.

The possibility of collective action is, however, further subverted
by Butler's assertion that the category 'woman' is itself unsustainable and only serves to limit feminist projects. The feminist subject is
generated by its representation. Gender identity is unstable because it
relies upon performativity which involves the repetition through acts
and gestures. Thus Butler has challenged the idea that gender is predicated upon sex; sex, gender and desire are all effects, that is none of
them are natural or essential. This has some purchase in an interrogation of the performativity of heteronormativity in sport, although
it is problematic in the political project of feminism if there is no
category women, not even one based on a strategic essentialism as
advocated by Gayatri Spivak.

Irigaray's politics of difference has been recuperated to address gender within a range of theoretical positions including the Deleuzian
(Braidotti, 2002). Irigaray's earlier work is in dialogue with and in
opposition to Lacanian psychoanalysis in asserting a non-essentialist
version of difference and presenting a challenge to the patriarchal

claims of the Oedipal complex and the Law of the Father. Contrary to Lacan and to Freudian universalism, she argues that Western culture rests on the death of the mother and that sexual difference is both embodied in its specificities and cultural in its manifestations and impact. Irigaray demonstrates the absence of the female pre-Oedipal imaginary and of representations of the mother daughter relationship and of the mother in Western thought, which offers a dynamic and powerful assertion of difference and of the female imaginary deriving from female corporeality (Irigaray, 1991, 1984). Irigaray transforms the relationship between bodies and culture in a politics of difference that, although sport was never part of her empirical brief, has enormous resonance with the politics of gender as played out in sport through bodies and discursive regimes, regulatory bodies and bodies of regulation.

Conclusion

Some of the examples highlighted in this chapter demonstrate the speed of change and the possibilities for transformation that are offered by technologies and new regulatory practices. It might be possible to move beyond bodies and to espouse a theory of becoming that opens up a radical approach to social and cultural change commensurate with the dynamic of technological progress. Conceptualizations of embodiment have to take account of the possibilities and sport is a site where such changes are powered by the economic, financial and cultural forces in play. Although sport is also characterized by an entrenched logic of differentiation and repeated attempts by regulatory bodies both to protect athletes from the excesses of the market and commercialization and to reestablish boundaries of difference, these attempts are deeply problematic. Instabilities abound and regulatory governance is subverted by enfleshed bodies that, for example, do not conform to the naturalized categories of gender adopted by regulatory bodies or of new technologies that create cyborgified humans who are difficult to situate. Technoscience both produces transformations and the means of engaging with embodied sporting practices by demonstrating some of the intersections of enfleshed corporeality and regulatory bodies.

Deleuzian thinking is useful for exploring non-linear change, progressions and mobilities but problematic in its lack of a politics of locations that can accommodate gender and difference in relation to the material situated body. The emphasis on change is very relevant to sport but sport also demonstrates stabilities, for example in its categorizations and practices, even in the face of technological advances, although this is a field in which change has been remarkably rapid. Disruptions, both to the linear narrative of change and to the notion of progress, focus on the enfleshed body in sport in particular ways and include the configuration of difference. Difference is important, as Braidotti argues, and an overemphasis on mobility, as within much Deleuzian philosophy, denies the place from which to speak and act as an embodied subject for those who are marked by gender difference. Gender remains an aspect of difference that is reiterated in regulatory practices and not fully accommodated by the theoretical positions that foreground the fluidity of bodies, whether enfleshed or regulatory. Change has been more frequently the focus of discursive systems than of approaches that address corporeality. Gender, however, is part of the assemblage of the embodied selves who are constituted through regulatory practices, the gendered narratives of sport and the enfleshed body that is both situated and itself a situation. Gender remains a useful aspect of difference, which, rather than being superseded in a post-gendered society is part of its constitution.

8
Conclusion

This book has focused upon the 'problem of bodies' and Butler's initial question about the relationship between material bodies and regulatory discursive regimes; regulating bodies and regulatory bodies. I have sought to move beyond the constraints of a polarization between social inscriptions and material corporeality, permitting a conceptualization of bodies and selves as interfaces for the assemblage of diverse energies and materialities. This book has covered some of the routes that have been taken in the development of this intellectual field, using the empirical vehicle of sport. The body has 'come back' theoretically in recent years as attempts have been made to reassert its materiality, although it has never, of course, been absent from the routine practices of sport, at all levels from the local gym or in the street to the biggest stadia that can accommodate thousands and feature celebrity athletes, or from its representations. Traditionally, in sport the body has been more a biological, anatomical concern, rather than part of the social and cultural theoretical analysis. Sporting bodies and sporting practices offer the possibility of transformation and of retrenchment and also highlight the disjunctions and disruptions as well as the continuities between disciplinary regimes and regulatory practices and the material bodies which they target. In a changing field that is both part of global economic forces and adapting regulatory practices and techniques, where body projects that reconfigure individuals through techniques of the self that include sport practices, sport has also become the motor for change and thus a site where the nature of change can be interrogated. This is particularly relevant in the context of theoretical shifts that emphasize the fluidity of change through the concept of assemblages characterized by multiplicities acting on 'semiotic flows, material flows and social flows simultaneously' (Deleuze and Guattari, 1987: 22). Without wanting to exaggerate the extent and fluidity of change, sport can be seen to make a useful contribution as an empirical site where transformations are taking place, sometimes at great speed, but where there are also continuities and traditional entrenchments. Change is uneven. Some of the points at which the speed of transformation might be expected to be the most dynamic, such as the widening of participation, for example, of women in sport at all levels, has met with the entrenchment of traditional binaries. Interventions are both advantageous and problematic. Sometimes they are both positive and negative simultaneously, as in the case of training regimes and pharmaceutical developments that themselves constitute a whole new fast progressing industry. Change is effected, both incrementally and in significant ways which can be understood as multiplicity, where change involves movement rather than a simple forward trajectory and different forces intersecting in the processes through which transformations are implicated.

The notion of an embodied or enfleshed subject is central to the discussion in this book and constitutive of the version of materialism which has been argued. The body has been presented as a conceptual space through which the tensions between nature and culture and the material and the discursive can be addressed from a range of analytic perspectives which were mapped out in Chapter 2. Most seek both to transcend binary logic and to in some way reinstate the enfleshed, material body, although, as I have argued, the body appears to have constituted a site saturated with discursive meanings and competing signs, where lived experience and the body as itself a situation have sometimes been obscured or marginalized. As the discussion in this book has shown, however, an engagement with the discursive regimes of governmentality through which embodied selves are regulated and disciplined does inevitably lead to a greater emphasis upon the social construction of meanings. The phenomenological development of theories of body practices, especially through the notion of the body as situation does provide a means of redressing some of this imbalance. Sport, in particular, highlights the limits of corporeality as well as its potential. The discussion moves from the tension between the discursive and the material towards understandings that elide different components of embodied subjectivity, rather than retaining a boundary between corporeality and discursive regimes or representations.

Phenomenological approaches provide a means of exploring the relationship between experience and knowledge and raise questions about how embodied knowledge is produced. An understanding of the situations in which knowledge is produced and how knowledge and the producers of knowledge are situated enables a fuller explanation of the diversity of the field. In sport this includes those who define the field as well as those who are embodied practitioners; the regulatory bodies and the regulated bodies and those who comment, which means those who research this field as well as the media and commentators of popular culture.

Sport is an empirical field which repeatedly appears to affirm polarizations of different sorts: between winners and losers, the natural and the fabricated, insiders and outsiders, fair play and fraud, cheating or deception and, especially, women and men. However, the field is more complex than is suggested by these configurations as was demonstrated in Chapter 3. Even what makes sport is contested. Debates about what constitutes sport have been concerned with corporeal, energetic bodies in play and in competition and questioned the extent to which what are seen as non-physical games so constitute sport. Competition is central to sport as configured in the contemporary world, although this is a characteristic that sits uneasily with the promotion of healthy minds and healthy bodies that forms a part of social inclusion and cohesion strategies of governance, as well as denying the pleasure in physical activity that is part of the lived body experience of sport for many people. The emphasis upon competition is often coded in through gender and configured around particular versions of masculinity. Sport is seen as universal, in that everyone who participates knows that there are rules that define a sport and that these apply more or less wherever the game is played but not everyone is included. Sport has a long history of racialized and ethnicized exclusion and the global sport media commerce nexus militates against equity internationally. The division between women's and men's sports remains firmly embedded, although the difficulty of establishing definitive guidelines on gender categories in some competitions, such as the Olympics, throws up questions about how boundaries are set and the centrality of theorizing difference in relation to embodiment. Science and technology are invoked, both to develop sports science, for example, through improved training regimes and to provide criteria for the evaluation of fair play in, for example, illegitimate performance enhancement through drugs or other interventions. Sport, in spite of its elaborate classificatory systems and cultural reinstatements of a binary logic, manifests many of the insecurities and uncertainties that are evidenced in contemporary culture. Embodied sporting practices also include activities that constitute resistance, whether within mainstream, traditional sports like boxing and cricket or in what have been called extreme sports.

The debate was moved more directly onto the question of inequality and difference in Chapter 4 in order to refocus on theories of governance at the macro-level and in particular to explore how the inequalities that permeate the field of sport can be both addressed and explained. Bodies on the margins present problems not the least in sport which is premised upon physical competence and competition. Sport is also widely enmeshed in commercial enterprise and part of global financial and economic networks, although classifications of embodied practices as sport that stress competition and reward can also exaggerate the notion that participation in sport necessarily means engaging either as a player or as a spectator in a particular range of traditional, mainstream sports only. The concept of diversity has been posited as a solution to the problem of equality and difference. The people who participate in sport and who follow sport are gendered and racialized; some occupy centre space and others are on the margins. People are classified by levels of dis/ability and specific body competences and, indeed, body parts. Difference, inequalities and marginalization are not only demarcated through body practices on the pitch or track or in the ring, sport is routinely experienced and represented and the two are inextricably linked; the axes of corporeality and body practices and the discursive regimes through which they are represented intersect and enmesh. Thus the media are central to the understanding what sport is. Sporting rules and regulations embody inequalities which are constitutive of what sport is and embodied differences are measured and quantified in sport through its regulatory mechanisms. One of the specificities of sport is that sport retains gendered distinctions which have been abandoned in other fields of experience and in other social relations. However, sport offers opportunities for inclusion and exclusion. There are contradictory identifications, where through the same apparatuses and practices, binaries, especially those of gender, are both retained and challenged. The politics of diversity has offered some of the opportunities of social inclusion in the neo-liberal project, but this is also a project that is deeply flawed in many a six respects. Neo-liberalism in its promotion of cohesion and social inclusion can be seen to operate conservatively in re-instating conformity; its aims are to stabilize communities, the wider society and the body politic. Activities within a framework of rights and translated into a politics of action offer more scope for change that could be transformative of lived experience of the embodied participants in sporting practices at whatever level of engagement.

Many aspects of marginalization in sport centre on the body. As was argued in Chapter 3, sport is concerned not only with body practices, but also with the enfleshed materiality of the embodied selves who participate in those practices; the dilemma with which the book began in mapping out contemporary theories of the body and embodiment. Foucauldian analyses of the apparatuses of governance may either subjectify the citizen self who is the target of the policies of inclusion and diversity as a body on which is inscribed by the regulatory discourses, or at least overemphasize how effective such mechanisms are at the level of the wider social terrain. There is limited concern with lived experience or with the micro-routine level of that experience. Power may operate diffusely but its description and explication operates most explicitly at the level of governance. The personal is political but the politics of the personal remain largely unexplored. Phenomenology presents useful focus upon actual bodies and such approaches have particular resonance within ethnographic studies of sport that permit reflection upon body practices and understanding through reflexive body practices.

Feminist phenomenological theories, as described in Chapter 4, in particular have pointed to the specificities of the gendered body and how gender is incorporated into these practices through a focus upon experience, although not very many have studied sport specifically. Iris Marion Young's approach to the gender-specific modalities of being in the world incorporates enfleshed practice with discursive regimes. There are, however, some problems with phenomenological approaches. First, although there is acknowledgement of the shared language and common institutional practices which act on and through the body, such approaches focus upon body comportment and individuals' perception of lived experience. There can be an overemphasis on experience of individuals at the expense of structural factors and the very significant inequalities that shape that experience. Sport is marked by significant inequalities which have to be addressed in explaining its meanings and relevance. Second, a methodological approach that is primarily concerned with recording lived experience and attributes agency to those whose experience is so recorded can involve an ontological complicity with the subjects in the field of research, as Merleau-Ponty admitted (1962). I have argued that feminist accounts, necessarily because feminism always has a political agenda and includes a critique of power structures however they operate, can address both of these criticisms. De Beauvoir's situated body is positioned and experienced in relation to the social, cultural and economic forces in play and these forces constitute the body as situation with its material, enfleshed actualities. However, the situated body and the body as situation do not entirely engage with how change can be effected beyond the idea of agency and the explanation for the intersection of social and embodied limitations and opportunities retains some dualisms that cannot fully explicate how difference and inequality intersect in the wider social terrain.

Regulatory bodies that occupy the discursive field and systems through which embodied selves are reproduced within sport include the media which play an enormously important role that extends beyond symbolic systems. Chapter 6 moved the debate beyond text to engage with some of the specificities of the media sport complex through spectacle, sensation and affect. Sport is spectacular as well as routine. Bodies are affected by sensation as practitioners and as spectators. Phenomenological accounts of sensation as the experience of the sensate body as situation combine the symbolic with the corporeal. From a very different but not entirely incompatible perspective the Deleuzian conceptualization of sensation as the event itself, and of affect as the body that affects and is affected, provides a radical departure from both linear narratives and the separation of the symbolic and the enfleshed. This is one way of transcending the separation between the flesh and the symbolic because there is no distinction and signification is undermined in the event of sensation which is distinguished from the sensational. The events at the

Bird's Nest Stadium in Beijing, at the Wacca or the Gabba in Australia, at Sabina Park in Jamaica, at Lords or Wimbledon in London, or Flushing Meadows in New York or at Madison Square Garden are themselves the affect. There is not a simple causal link between the show that is staged and the performance of sporting super stars in the arena and the audience of fans who react as spectators. The Super Bowl at Tampa in 2009 was hailed as a great show because the fans, the players, the game and the place were all implicated in the same event and sensation. However, when thinking about sporting mega events, it is hard not to comment on the spectacle as sensation, since so much contemporary global sport is configured within the rhetoric of drama and pivotal moments and expressed in the hyperbole of mediatized sensational news.

Media representations of sport are, however, sensational, especially in the alliance between sport and popular culture and the configuration of the sport star or celebrity. Both celebrities and heroes are recreated through the trope of narrative, which plays no part in the Deleuzian version of sensation. Its appeal is hard to deny though and the sequential construction of the story presents a powerful means of the embedding of masculinity in the meanings attached to and experienced by the sporting hero or the celebrity. I have argued, however, that there are distinctions that can be made between the hero and the celebrity through an analysis of the media sport relationship which itself does not follow a simple or linear narrative. Stories, especially in sport, are most commonly vehicles for the reproduction of a male hero and a masculine central protagonist, but this in itself does not eliminate the importance of narrative which is a force in the assemblages through which embodied sporting selves are made and remade.

Postmodernist theorizing on bodies and embodiment has focused upon the transcending of boundaries that move beyond that between the anatomical body and the discursive regimes which impact upon it. The subversion of inequalities has been posed through the promise of monsters and especially through the possibilities of cyborgification and becoming non-human; the borders between the human and the animal and between human and machine blur in what could be productive entanglement, especially for those who have been most marginalized or devalued by the category of human. Sport may be sensational, but it is also transforming and transformational. One motor for change is one of the most sensational and widely imbricated with sporting bodies. To achieve sporting excellence there is the rigorous routine of training, nutritional programmes and a complex relationship with the advances of science and technology. When technoscience is within the boundaries of legality it is legitimate but progress is often so fast that these boundaries are blurred and performance enhancement extends beyond that which is permitted. Regulatory bodies create categories of athletes; there are the Olympics and the Paralympics, but who can be included in which Games bring together the regulatory body and the regulated body within the athlete. As I have repeated through this book, sport endeavours to hold onto its own ideas about gender difference, but the relationship between science and embodiment is disruptive and the uses to which science can be put are enmeshed with social and ethical systems. Sport is part of a wider terrain in which networks that reinstate collusive masculinities and racialized connections circulate and, because of the centrality of corporeality and material bodies and body practices in sport, it is a site at which such networks, for example of hegemonic masculinity, retain considerable influence. The enfleshed bodies of athletes are the site at which these different power axes intersect. Sport produces its own cyborgs although the promise transformation and liberation cannot always be delivered. Boundaries shift but are renegotiated at the interface between the discursive and the material and corporeal.

Embodied sporting practices invoke different dimensions of the forces which create them. Regulatory bodies reproduce and create body practices which are transformed in the process. These practices are diverse and are not confined to the experience of participating in sport but encompass spectatorship and a whole range of experiences that are connected to what is defined as sport but might be in tangential ways. This is not a straightforward or linear trajectory and one of the most useful aspects of Deleuzian assemblages is the possibility of disruption. An assemblage establishes connections with previously separate entities reuniting them into a complex whole. This emphasis on mobility challenges the fixity of oppositions, which are nonetheless deeply embedded in the discourse of sport and in the voices of its participants. Everything is mixed up in assemblages that suggest an interaction and intertwining of forces, not the determination of one force by another. However, Deleuzian assemblages are based on the systems which inform them. Whatever the radical critique of philosophy which Deleuze proposes, assemblages of intersecting systems seem far removed from lived experience especially of those on the margins, and present limited scope for the voices of those who are excluded. Given the centrality of difference in sport, this book throws up questions about how effectively this version of an assemblage can deal with the materialities of difference and the embodied self who is also the body as situation. The concept of 'becoming' while useful in engaging with mobilities and understanding change is less so when it comes to recognizing and addressing difference, which is part of being. Sport can be seen to provide assemblages that include the articulation of the corporeal and discursive elements of a given assemblage. The elements of assemblage change, and when they do so they do practice and experience, which also allows some possibility of intervention and of productive transformation. Change can take place if sport is seen as an assemblage, which includes the intersection of different axes of power including different flows of economic, social and political power *and* the body practices of those who participate and the discursive dimensions through which they are represented, in which the media play a key role.

The notion of bodies as situated and in situations can incorporate the body as a set of forces also offers a useful way of avoiding the limitations of the binary logic or the discursive and corporeal and even of suggesting that such a dualism may be the wrong starting point. Disjunction and disruption are also constitutive of embodied sporting practices and as is most sensationally demonstrated in sport, broken bodies can evidence the body fighting back in a sensational and sentient manner. Sport is marked by routines and by extremes; the rigorous training regimes, the success of the mega events and local triumphs and the disastrous failures and routine injuries all combine to make the body as situation central to the project. Bodies as situations encompass all participants, including the spectator, the researcher and the commentator, which also challenges one of the most entrenched binaries even in research that addresses reflexivity and the need to engage with the relationship between the subjects of research and the researcher. The academy is not only a regulatory body that is complicit in the production of meanings about embodied sporting practice, its members are also situated within the field and themselves bodies as situations.

Butler has argued that an adequate theory has to reinstate the body and find a means of breaking down the barriers between the enfleshed body and the discursive field and to provide a means to showing that bodies matter without ending up only able to explain difference in relation to the anatomical body. Feminist theorists have been strongly motivated to engage with this problem, although, however mobile and fluid some approaches appear, they might have been addressing the wrong problem. The idea of reinstating the body suggests that it is a separate entity outside the discursive field and retains some notion of separation. The alternative is not only to claim that the body is socially constructed, however complex the processes, but to recognize the entanglement of the corporeal and the discursive without privileging one over the other. These inextricable connections are forged temporally and spatially, but in different ways at different times. Difference is not only marked in embodiment, but it also characterizes moments of change and materialities that are inflected by power geometries that themselves transform and mutate. Bodies matter, to some extent, all the time but they matter more at some points than others, just as difference is pivotal in some contexts and should not be in others. Feminists have often sought to put sexuality and sex/gender, especially gender into the debate about bodies whilst avoiding the threat of biology. Criticisms that this might have led to an excessive focus upon sexual/gendered difference are less likely to apply in sport, where the concern with gender is part of lived experience of those who are involved. A focus on gender and difference makes situations explicit and opens up the possibility of interrogating masculinity and male bodies too. Although sports studies have recently taken up what have been called pro-feminist approaches and have interrogated masculinities, there are still significant silences, notably within the collusions of embodied sporting practices off the pitch as much as on. Crude binary logic cannot explain difference. Situatedness also permits the inclusion of gender without asserting that it is the only difference as well as acknowledging the multiplicities of differentiation. The symbolic is also a component in the situated body in its situation and not something extraneous and outside the routine practices of embodiment. Bodies have effects and affects which are manifest in the contingent intersections of forces that are also differentially weighted and inflected. These intersections always include the regulatory bodies and the discursive and the embodied practices. Thus, embodied sporting practices can be located within the concept of an assemblage if that assemblage incorporates corporeality and the body as situations as well as the discursive systems though which meanings and differences are reproduced and the dynamics of potential transformations are always in play. The problem of bodies is not a problem if bodies are construed as material enfleshed and disciplined and inscribed. To engage with enfleshed materialities is not to reduce subjects to their corporeality, nor to suggest that changes are not possible and imaginable; transformations become credible and possible through being attentive to lived embodied experience and embodied sporting practices are a productive field of enquiry to this end. Change is possible, even, perhaps especially, in sport, through disruption as well as through progress and is not always predictable. Sometimes mobilities can be grasped through exploring what remains and, especially through what is material.

Bibliography

Adebayor, E. (2008) http://www.bbc.co.uk/pressoffice/pressreleases/stories/2008/05_may/23/adebayor.shtml (last accessed 6 April 2009).

Adlington (2008) http://news.bbc.co.uk/sport1/hi/tv_and_radio/sports_personality_of_the_year/7733581.stm (last accessed 3 February 2009).

Ali, M. and Durham, R. (1975) *The Greatest: My Own Story*. New York: Random House.

Anderson, B. (1983) *Imagined Communities: Reflections on the Origins and Spread of Nationalism*. London: Verso.

Anderson, B. (1991) *Imagined Communities*, revised edition. London: Verso.

Andrews, D. (2006) *Sport-Commerce-Culture*. New York: Peter Lang.

Andrews, D. and Jackson, S. (eds) (2001) *Sport Stars: The Politics of Sporting Celebrity*. London: Routledge.

Armstrong, G. and Giulianotti, R. (1999) *Football, Culture and Identities*. Basingstoke: Macmillan.

Asians in Football Forum (2005) *Asians Can Play Football: Another Wasted Decade*. London: Asians in Football Forum.

Atkinson, M. (2008) *Battleground: Sports*. Westport: Greenwood Press.

Back, L., Crabbe, T. and Solomos, J. (1998) *Racism in Football*. Basingstoke: Macmillan.

Back, L., Crabbe, T. and Solomos, J. (2001) *The Changing Face of Football Racism, Identity and Multiculture in the English Game*. Oxford: Berg.

Bakhtin, M. (1984) *Rabelais and His World*. Bloomington: Indiana University Press.

Bale, J. (2000) 'Sport as Power', in Sharp, J.P., Routledge, P., Philo, C. and Paddison, R. (eds) *Entanglements in Power*. London: Routledge.

Balsamo, A. (1996) *Technologies of the Gendered Body*. Durham: Duke University Press.

BBC 606 (2008) http://www.bbc.co.uk/dna/606/A36072308 (last accessed 3 November 2008).

BBC Press Release (2008) http://www.bbc.co.uk/pressoffice/pressreleases/stories/2008/05_may/23/adebayor.shtml (last accessed 6 April 2009).

Battersby, C. (1998) *The Phenomenal Woman*. Cambridge: Polity.

Barry, B. (2001) 'The Muddles of Multi Culturalism', *New Left Review*, 8, pp. 49–71.

Beattie, G. (1996) *On the Ropes: Boxing as a Way of Life*. London: Indigo Cassell.

Beauvoir, S. de ([1949] 1987) *The Second Sex*, trans. H. Parshley. New York: Vintage Books.

Beauvoir, S. de (1989 [1949]) *The Second Sex*. London: Vintage Books, trans. H. Parshley from *Le Deuxieme Sexe*, 2 vols, Paris: Gallimard.

Bell, M. (2003) 'The Right to Equality and Non-Discrimination', in Hervey, T. and Kenner, J. (eds) *Economic and Social Rights under the EU Charter of Fundamental Rights: A Legal Perspective*. Oxford: Hart Publishing, pp. 91–110.

Blackman, L. (2008a) 'Affect and Feeling', *Critical Psychology: International Journal of Critical Psychology Theory, Culture & Society*, 1 January, 25(1), pp. 23–47.

Blackman, L. (2007) *Psychiatric Culture and Bodies of Resistance Body & Society*, 13(2), pp. 1–23.

Blackman, L. and Cromby, J. (2007) 'Affect and Feeling', *International Journal of Critical Psychology*, 21(1), pp. 5–22.

Blackman, L. (2008b) *Affect and Relationality and the 'problem of personality' Theory Culture and Society*, 25(1), pp. 23–47.

Blalock, H. (1962) 'Occupational Discrimination', *Social Problems*, 9, pp. 240–247.

Blalock, H. (1967) *Toward a Theory of Minority-Group Relations*. New York: John Wiley.

Block, D. (2005) *Baseball Before We Knew It: A Search for the Roots of the Game*. Nebraska: University of Nebraska Press.

Boorstin, D. (1992) *The Image: A Guide to Pseudo-Events in America*. New York: Random House.

Bordo, S. (1993) *Unbearable Weight: Feminism, Western Culture, and the Body*. Berkeley: University of California Press.

Born, G. (2005) *Uncertain Visions: Birt Dyke and the Reinvention of the BBC*. London: Vintage.

The Boston Women's Health Collective (1978 [1971]) *Our Bodies Ourselves*. Harmondsworth: Penguin Books UK edition, First published in the US 1971.

Bourdieu, P. (1977) *Outline of a Theory of Practice*. Cambridge: Cambridge University Press.

Bourdieu, P. (1984) *Distinction: A Social Critique of the Judgement of Taste*, trans. R. Nice. Cambridge, MA: Harvard University Press.

Bourdieu, P. ([1997] 2000) *Pascalian Meditations*, trans. L. Wacquant. Cambridge: Polity Press.

Bourdieu, P. (1992) *The Logic of Practice*. Cambridge: Polity.

Bourdieu, P. (2000) *Pascalian Meditations*. London: Routledge.

Bradbury, S. (2001) *Football Unites, Racism Divides. An Evaluation of the Period 1998–2000*. University of Leicester, Sir Norman Chester Centre for Football Research.

Braidotti, R. (1994) *Nomadic Subjects: Embodiment and Sexual Difference in Contemporary Feminist Theory*. New York: Columbia University Press.

Braidotti, R. (1996) 'Signs of Wonder and Traces of Doubt: On Teratology and Embodied Differences', in Lykke, N. and Braidotti, R. (eds) *Between Monsters, Goddesses and Cyborgs*. London: Zed Books.

Braidotti, R. (1991) *Patterns of Dissonance*. Cambridge: Polity.

Braidotti, R. (2002) *Metamorphoses: Towards a Materialist Theory of Becoming*. Cambridge: Polity.

Brearley, M. (2009) The Tiger takes Guard, BBC Radio 4, 8 p.m. 2 February.

Breitenbach, E. (ed.) (2002) *The Changing Politics of Gender Equality in Britain*. Basingstoke: Palgrave Macmillan.

Brown, W. (2003) 'Neo-liberalism and the End of Liberal Democracy', *Theory and Event*, 7(1), pp. 4–25.

Brown, A., Crabbe, T., Mellor, G., Blackshaw, T. and Stone, C. (2006) *Football and its Communities: Final Report*. Football Foundation and Manchester Metropolitan University, London and Manchester.

Buchanan, I. and Coleman, C. (eds) (2000) *Deleuze and Feminist Theory*. Edinburgh: Edinburgh University Press.

Burchell, G., Gordon, C. and Miller, P. (eds) (1991a) *The Foucault Effect: Studies in Governmentality*. London: Harvester Wheatsheaf.

Burchell, G., Gordon, C. and Miller, P. (1991b) *The Foucault Effect*. Chicago Illinois: University of Chicago Press.

Burdsey, D. (2007) *British Asians and Football: Culture, Identity and Exclusion*. London: Routledge.

Butler, J. (1993) *Bodies That Matter, On the Discursive Limits of Sex*. London: Routledge.

Butler, J. (1990) *Gender Trouble: Feminism and the Subversion of Identity*. London: Routledge.

Butler, J. (1997) *Excitable Speech: A Politics of the Performative*. New York: Routledge.

Canadian Press (2009) http://www.google.com/hostednews/canadianpress/article/ALeqM5i48bxrtVHJpNkwiuz8z269tHEvBQ (last accessed 31 January).

Cardus, N. (1947) *English Cricket*. London: Collins.

Cardus, N. (1970) *Full Score*. London: Collins.

Cardus, N. and Arlott, J. (1986) *The Noblest Game: A Book of Fine Cricket Prints*. London: Bloomsbury.

Carrington, B. (1998) 'Sport Masculinity and Black Cultural Resistance', *Journal of Sport and Social Issues*, 33(3), pp. 275–298.

Carrington, B. (2001) 'Introduction: "Race", Sport and British Society', in Carrington, B. and McDonald, I. (eds) *Race, Sport and British Society*. London: Routledge, pp. 1–26.

Carrington, B. and McDonald, I. (eds) (2001) *'Race', Sport and British Society*. London: Routledge.

Carrington, B. and McDonald, I. (2008) 'The Politics of "Race" and Sports Policy in the United Kingdom', in Houlihan, B. (ed.) *Sport and Society*, 2nd edition. London: Sage, Chapter 10, pp. 230–254.

Cary, J. (2005) 'Spectacle', in Bennett, T., Grossberg, L. and Morris, M. (eds) *New Keywords: A Vocabulary of Culture and Society*. Oxford: Blackwell, pp. 335–336.

Cashmore, E. (2004) *Celebrity Culture*. London: Routledge.

Cashmore, E. (2005) *Making Sense of Sport*, 4th edition. London: Routledge.

Charlton Athletic (2007), http://www.cafc.co.uk/yourview.ink?messageid=15421&display (last accessed 16 October 2007).

Chick, G. (2004) 'History of Games', in McNeil, W.H. (ed.) *Berkshire Encyclopaedia of World History*. Great Barrington, MA: Berkshire Publishing, pp. 798–802.

Clough, P.T. with Halley, J. (eds) (2007) *The Affective Turn: Theorizing the Social*. Durham, NC: Duke University Press.

Cole, C. (1998) 'Addiction, Excess and Cyborgs', in Rail, C. (ed.) *Sport in Post-Modern Times*. New York: New York University Press.

Coles, R. (2005) *Beyond Gated Communities: Reflections for the Possibility of Democracy*. Minneapolis: University of Minnesota Press.

Collins, M., Henry, I., Houlihan, B. and Buller, J. (1999) *Research Report: Sport and Social Exclusion*. Loughborough: Loughborough University.

Connell, R.W. (1995) *Masculinities*. Cambridge: Polity.

Connell, R.W. (2002) *Gender*. Cambridge: Polity.

Connell, R.W. (2005) 'Change among the Gatekeepers: Men, Masculinities and Gender. Equality in the Global Arena', *Signs: Journal of Women in Culture and Society*, 30(31), pp. 1801–1825.

Crabbe, T. (2004) 'Englandfans a New Club for a New England? Social Inclusion, Authenticity and the Performance of Englishness at "home" and "away" ', *Leisure Studies*, 23(1), pp. 63–78.

Crabbe, T. and Brown, A. (2004) ' "You're not welcome anymore": The Football Crowd, Class and Social Exclusion', in Wagg, S. (ed.) *British Football and Social Exclusion*. London: Routledge.

Crawford, G. (2003) 'The Career of the Sport Supporter: The Case of the Manchester Storm', *Sociology*, 37(2), pp. 219–237.

Crawford, G. (2004) *Consuming Sport*. London: Routledge.

Crossley, N. (2001) *The Social Body: Habit, Identity and Desire*. London: Sage.

Caudwell, J. (2003) 'Sporting Gender: Women's Footballing Bodies as Sites/Sights for the (re) Articulation of Sex Gender', *Sociology of Sport Journal*, 20(4), pp. 371–386.

Caudwell, J. (2007) 'Sex and Politics: Sites of Resistance in Women's Football', in Tomlinson, A. (ed.) *The Sports Studies Reader*. London: Routledge, pp. 356–359.

Daly, M. (1978) *Gyn/Ecology: The Metaethics of Radical Feminism*. Beacon: Beacon Press.

DCMS (2000) *A Sporting Future for All*. London: HMSO.

DCMS (2001) *The Government's Plan for Sport*. London: HMSO.

DCMS/SU (2002) *Game Plan: A Strategy for Delivering Government's Sport and Physical Activity Objectives*. London: Cabinet Office.

Debord, G. (1990) *Comments on the Society of the Spectacle*, trans. M. Imrie. London: Verso.

Debord, G. (1994) *The Society of the Spectacle*, trans. D. Nicholson-Smith. New York: Zone Books.

De Landa, M. (2006) *A New Philosophy of Society: Assemblage and Social Complexity*. London and New York: Continuum.

Deleuze, G. (1994) *Difference and Repetition*, trans. P. Patton. New York: Columbia University Press.

Deleuze, G. (2005) *Francis Bacon*, trans. Daniel W. Smith. London and New York: Continuum.

Deleuze, G. and Guttari, F. (1987) *A Thousand Plateaus: Capitalism and Schizophrenia*, trans. B. Massumi. London: Athlone Press.

de Garis, L. (2000) 'Be a Buddy to Your Buddy', in McKay, J., Messner, M. and Sabo, D. (eds) *Masculinities, Gender Relations and Sport*. London: Sage.

Derrida, J. (1994) *Specters of Marx: The State of Debt, the Work of Mourning and the New International*, trans. P. Kamuf. New York: Routledge.

Doane, A.W. (2006) 'What is Racism? Racial Discourses and Racial Politics', *Critical Sociology*, 32(2–3), pp. 255–274.

Doane, M.A. (1999 [1990]) 'Technophilia, Technology, Representation and the Feminine', in Kirkup, G., Janes, L., Woodward, K. and Hovendon, F. (eds) *The Gendered Cyborg*. London: Routledge, pp. 110–121.

Donnelly, P. (ed.) (1996) *Taking Sport Seriously: Social Issues in Canadian Sport*, 2nd edition, 2000. Toronto: Thompson Educational Publishing, p. 248.

Donzelot, J. (1980) *The Policing of Families*. London: Hutchinson.

Douglas, M. (1966) *Purity and Danger: An Analysis of the Concepts of Pollution and Taboo*. London: Routledge.

Dower, J. (2008) *True Stories: Thrilla in Manila*, More 4, 12 November.

Dutton, K. (1995) *The Perfectible Body: The Western Ideal of Physical Development*. London: Cassell.

Dunning, E. (1999) *Sport Matters: Sociological Studies of Sport and Leisure and Civilization*. London: Routledge.

Dunning, E. and Sheard, K. (1979) *Barbarians, Gentlemen and Players: A Sociological Study of Rugby Football*. Oxford: Robertson.

Early, G. (1994) *The Culture of Bruising: Essays on Prizefighting, Literature and Modern American Culture*. Hopewell, NJ: ECCO.

Early, G. (1998) 'Performance and Reality: Race, Sports and the Modern World', *The Nation* (10–17 August), pp. 11–20.

Elias, N. (1978) *The Civilizing Process*. Oxford: Blackwell.

Elias, N. (1982) *A History of Manners*. London: Random House.

Elias, N. and Dunning, E. (1986) *Quest for Excitement: Sport and Leisure in the Civilizing Process*. Oxford: Basil Blackwell.

Egan, P. (1812) *Boxiana; or Sketches of Ancient and Modern Pugilism; from the Days of the Renowned Broughton and Slack to the Heroes of the Present Milling Era*. London: Sherwood.

FA Women (2007) http://www.thefa.com/Womens/ (last accessed 6 April 2009).

FA Women (2009) http://www.thefa.com/Womens/GettingInvolved/ (last accessed 31 January 2009).

FARE (2008) http://www.farenet.org (last accessed 6 January 2008).

Fast track (2008) http://bbc.fasttrack (last accessed 4 December 2008).

Fausto-Sterling, A. (2000) 'Sexing the Body', *Gender Politics and the Construction of Sexuality*. New York: Basic Books, Perseus Books Group.

Featherstone, M. (ed.) (1999) 'Body Modification', Special issue of *Body and Society*, 5(2/3), pp. 1–13.

Featherstone, M. and Hepworth, M. (eds) (1991) *The Body: Social Process and Cultural Theory*. London: Sage.

FIFA (2006) http://www.fifa.com/aboutfifa/media/newsid=529882.html (last accessed 31 January 2009).

Fincher (1999) *Fight Club*.

Football Unites, Racism Divides (2006) *The First Ten Years, 1995–2005*. Sheffield: Stables Connexion Centre.

Foucault, M. (1973a) *Madness and Civilization*. New York: Vintage Books.

Foucault, M. (1973b) *The Order of Things*. New York: Vintage Books.

Foucault, M. (1974) *The Archaeology of Knowledge*. London: Tavistock.

Foucault, M. (1977) *Discipline and Punish*. New York: Pantheon Books.

Foucault, M. (1980) *Power/Knowledge*. New York: Pantheon Books.

Foucault, M. (1981) *The History of Sexuality. Volume 1: An Introduction*, trans. R. Hurley. Harmondsworth: Penguin.

Foucault, M. (1982) 'The Subject and Power', in Dreyfus, H. and Rabinow, P. (eds) *Michel Foucault: Beyond Structuralism and Hermeneutics*. Chicago: University of Chicago Press.

Foucault, M. (1988a) *The History of Sexuality. Volume 2: The Use of Pleasure*. New York: Vintage Books.

Foucault, M. (1988b) *The History of Sexuality. Volume 3*. New York: Vintage Books.

Foucault, M. (2001) 'Governmentality', in Faubion, J.D. (ed.) *Power. Essential Works 1, 1954–84*. Harmondsworth: Penguin, pp. 201–222.

Fozooni, B. (2008) 'Iranian Women and Football', *Cultural Studies*, 22(1), pp. 114–133.

Fraser, M. and Greco, M. (eds) (2005) *The Body: A Reader*. London: Routledge.

Free, M. and Hughson, J. (2003) 'Settling Accounts with Hooligans: Gender Blindness in Football Supporter Sub Culture Research', *Men and Masculinities*, 6(2), pp. 136–155.

Frosh, S. (2002) *Afterwords*. Basingstoke: Palgrave Macmillan.

Gamson, J. (1994) *Claims to Fame: Celebrity in Contemporary America*. Berkeley: University of California Press.

Gatens, M. (1996) *Imaginary Bodies: Ethics, Power and Corporeality*. London: Routledge.

Gayo-Cal, M., Savage, M. and Warde, A. (2006) 'A Cultural Map of the United Kingdom 2003', *Cultural Trends*, 15(2/3), pp. 213–237.

Genel, M. (2000) 'Gender Verification No More?', *Medscape Women's Health*, 5(3) Medscape, Inc, http://ai.eecs.umich.edu/people/conway/TS/Olympic-GenderTesting.html (accessed 7 December 2008).

Gilroy, P. (2005) *Postcolonial Melancholia*. New York: Columbia University Press.

Giulianotti, R. (1991) 'Scotland's tartan Army in Italy', *Sociological Review*, 39(3), pp. 503–527.

Giulianotti, R. (1995) 'Football and the Politics of Carnival', *International Review for the Sociology of Sport*, 30(2), pp. 191–224.

Giulianotti, R. (1999) *More Than a Game: The Social and Historical Aspects of World Football*. Cambridge: Polity.

Giulianotti, R. (2005a) *Sport a Critical Sociology*. Cambridge: Polity.

Giulianotti, R. (2005b) 'Sport Spectators and the Social Consequences of Com-modifcation, Critical Perspectives from Scottish Football', *Journal of Sport and Social Issues*, 29(4), pp. 386–410.

Giulianotti, R. (2005c) *Sport and Modern Social Theorists*. London: Palgrave Macmillan.

Giulianotti, R. and Williams, J. (eds) (1994) *Game Without Frontiers: Football, Modernity, Identity*. Aldershot: Arena.

Giroux, H. (2005) *The Terror of Neoliberalism*. New York: Palgrave Macmillan.

Goffman, E. (1959) *The Presentation of Self in Everyday Life*. Garden City, NY: Doubleday Anchor.

Goldblatt, D. (2006) *The Ball is Round: A Global History of World History of Football*. London: Penguin.

Gorn, E.J. (1986) *The Manly Art: The Lives and Times of the Great Bare Knuckle Champions*. London: Robson Books.

Gratton, C. and Taylor, P. (2004) *The Economics of Sport and Recreation*. London: Routledge.

Grieveson, L. (1998) 'Fighting Films', *Cinema Journal*, 38(1), pp. 40–72.

Grossberg, L. (2005) *Caught in the Crossfire: Kids, Politics and America's Future*. Boulder CO and London: Paradigm.

Grosz, E. (1990) 'The Body of Signification', in Fletcher, J. and Benjamin, A. (eds) *Abjection, Melancholia and Love*. London: Routledge.

Grosz, E. (1994) *Volatile Bodies: Toward a Corporeal Feminism*. Bloomington, IN: Indiana State University.

Grosz, E. (1995) *Space, Time and Perversion: Essays in the Politics of Bodies*. London and New York: Routledge.

Grosz, E. (1999) 'Psychoanalysis and the Body', in Price, J. and Shjldrick, M. (eds) *Feminist Theory and the Body*. Edinburgh: Edinburgh University Press.

Grosz, E. (2000) 'Deleuze's Bergson: Duration, the Virtual and a Politics of the Future', in Buchanan, I. and Colebrook, C. (eds) *Deleuze and Feminist Theory*. Edinburgh: Edinburgh University Press.

Gruneau, R. (1993) 'The Critique of Sport in Modernity: Theorizing Power, Culture, and the Politics of the Body', in Dunning, E.G., Maguire, J.A. and Pearton, R.E. (eds) *The Sports Process: A Comparative and Developmental Approach*. Champaign, IL: Human Kinetics, pp. 85–110.

Grupe, O. (2005) *Vom Sinn des Sports* (Schorndorf: Karl Hoffman), cited by Allen Guttmann, p. 328.

Guttmann, A. (1978) *From Ritual to Record*. New York: Columbia University Press.

Guttmann, A. (2002) *The Olympics: A History of the Modern Games*, 2nd edition. Champaign, IL: University of Illinois Press.

Guttmann, A. (2005) *Sport the First Five Millennia*. Amherst and Boston: University of Massachusetts Press.

Hall, M.A. (1996) *Feminism and Sporting Bodies*. Champaign, IL: Human Kinetics.

Hall, S. (2001) 'Conclusion to the Multi-Cultural Question', in Hesse, B. (ed.) *Un/Settled Multiculturalisms*. London: Zed Books, pp. 209–241.

Hammersley, M. and Atkinson, P. (1995) *What's Wrong With Ethnography: Methodological Explorations*. London: Routledge.

Hammersley, M. and Atkinson, P. (2005) *Ethnography: Principles in Practice*, 2nd edition. London: Routledge.

Haraway, D.J. (1990) *Primate Visions: Gender, Race, and Nature in the World of Modern Science*. London and New York: Routledge.

Haraway, D. (1991) *Simians, Cyborgs and Women: The Reinvention of Nature*. London: Free Association Books.

Haraway, D. (1996) *Modest Witness@Second Millennium: FemaleMan Meets OncoMouseTM: Feminism in the Late Twentieth Century*. New York: Routledge.

Haraway, D. (2007) *When Species Meet*. Minneapolis: University of Minnesota Press.

Haraway, D. (2008) *The Haraway Reader*. London and New York: Taylor and Francis.

Hargreaves, J. (1986) 'The Autonomy of Sport, Sport', *Power and Culture: A Social and Historical Analysis of Popular Sports in Britain*. Cambridge: Polity, pp. 75–86.

Hargreaves, J. (1987) 'Sport, Body and Power Relations', in Horne, J., Jery, D., and Tomlinson, A. (eds) *Sport, Leisure and Social Relations*. London: Routledge and Kegan Paul.

Hargreaves, J. (1994) *Sporting Females*. London: Routledge.

Hargreaves, J. (2000) 'Querying Sport Feminism: Personal or Political?, in Giulianotti, R. (ed.) *Sport and Modern Social Theorists*. Basingstoke: Palgrave Macmillan, pp. 187–206.

Hargreaves, J. (2004) 'Querying Sport Feminism: Personal or Political', in Giulianotti, R. (ed.) *Sport and Modern Social Theorists*. London: Palgrave Macmillan, pp. 187–206.

Hargreaves, J., Vertinsky, P. and McDonald, I. (eds) (2007) *Physical Culture, Power and the Body*. London: Taylor and Francis.

Hart, S. (2008) 'Pistorius Tramples Over Notions of Equality', *London, Sport Telegraph*, 9 September 2008, p. S15.

Hauser, T. (1991) *Muhammad Ali: His Life and Times*. New York: Simon Schuster.

Hauser, T. (2007) 'The Brown Bomber is destroyed in his last ever fight': London, *Observer Sport Monthly*, p. 53.

Hazlitt, W. (1982) 'The Fight', *William Hazlitt, Selected Writings*. Harmondsworth: Penguin.

Heater, D. (1999) *What is Citizenship*. Cambridge: Polity Press.

Holt, R. and Mason, T. (2007) 'Sensationalism and the Popular Press', in Tomlinson, A. (ed.) *The Sports Studies Reader*. London: Routledge.

Home Office (2005) *Improving Opportunity, Strengthening Society: The Government's Strategy to Increase Race Equality and Community Cohesion*. London: Race Cohesion, Equality and Faith Directorate, Crown Copyright.

Honea, J. (2004) 'Youth Cultures and Consumerism: Sport Subcultures and Possibilities for Resistance', PhD Dissertation, Colorado State University, Fort Collins.

Houlihan, B. (ed.) (2008) *Sport and Society*, 2nd edition. London: Sage.

Howe, D. (2006) 'Tebbit's Loyalty Test is Dead', *New Statesman*, 3 July, London.

Howson, A. (2005) *Embodying Gender*. London: Sage.

Huizinga, J. (1955) *Homo Ludens: A Study of the Play-Element in Culture*. Boston: Beacon Press.

Husserl, E. (1970) *The Crisis of European Sciences and Transcendental Phenomenology: An Introduction to Phenomenological Philosophy*, trans. D. Carr. Evanston: North Western University Press.

Ilde, D. (2009) If phenomenology is an albatross, is *postphenomenology* possible? http://www.sunysb.edu/philosophy/faculty/dihde/articles/postphenomenology.html (last accessed 6 April 2009).

Ingham, A.G. (1978) 'American Sport in Transition: The Maturation of Industrial Capitalism and Its Impact on Sport', PhD dissertation, University of Massachusetts, Amherst.

Ingham, A.G. (2004) 'The Sportification Process: A Biographical Analysis Framed in the Work of Marx [Engels], Weber, Durkheim and Freud', in Giulianotti, R. (ed.) *Sport and Modern Social Theorists*. Basingstoke: Palgrave Macmillan, pp. 11–32.

Ingham, A.G. and Loy, J.W. (1973) 'The Social System of Sport: A Humanistic Perspective', *Quest*, 1(19), pp. 3–23.

Ingham, A.G. and Loy, J.W. (eds) (1993) *Sport in Social Development: Traditions, Transitions, and Transformations*. Champaign, IL: Human Kinetics.

IOC (2009) http://www.olympic.org/uk/index_uk.asp (last accessed 6 April 2009).

IPL South Africa (2009) http://www.independent.co.uk/sport/cricket/ipl-to-be-staged-in-south-africa-1653027.html (last accessed 12 May 2009).

IPL (2008) http://www.thetwenty20cup.co.uk/db/indian_premier_league/default.asp (last accessed 6 April 2009).

Irigaray, L. (1984) *Speculum of the Other Woman*, trans. G. Gill. Ithaca, NY: Cornell University Press.

Irigaray, L. (1989) 'Sexual Difference', in Moi, T. (ed.) *French Feminist Thought*. Oxford: Basil Blackwell.

Irigaray, L. (1991) 'This Sex Which is Not One', in Whitford, M. (ed.) *The Irigaray Reader*. Oxford: Basil Blackwell.

James, C.L.R. (1963) *Beyond a Boundary*. London: Stanley Paul.

Jeu, B. (1972) *Le Sport, La Mort, La Violence*. Paris: Éditions Universitaires.

Joppke, C. (2004) 'Ethnic Diversity and the State', *British Journal of Sociology*, 55(3), pp. 451–463.

Kay, S. (2003) *Zizek. A Critical Introduction*. Cambridge: Polity.

Kick it Out (2007) www.kickitout.org (last accessed 7 October 2007).

Kick it Out (2008) www.kickitout.org.uk (last accessed 20 December 2008).

Klapp, O. (1949) 'Hero Worship in America', *American Sociological Review*, 14(1), pp. 57–63.

Klein, A. (1993) *Little Big Men: Bodybuilding Subculture and Gender Construction*. Albany: State University of New York Press.

Knight, T. (2008) New drugs Test Cast Shadow Over Olympics, 14 October 2008, http://www.telegraph.co.uk/sport/othersports/drugsinsport/3197570/

New-drugs-tests-cast-shadow-over-Olympics-Athletics.html (last accessed 7 December 2008).

Kram, M. (2002) *Ghosts of Manila*. London: Harper Collins.

Kristeva, J. (1987) *The Kristeva Reader*, Moi, T. (ed.). Oxford: Basil Blackwell.

Lafferty, Y. and McKay, J. (2005) ' "Suffragettes in Satin Shorts?" Gender and Competitive Boxing', *Qualitative Sociology*, 27(3), pp. 249–276.

Lahore (2009) http://news.bbc.co.uk/1/hi/world/south_asia/7921291.stm (accessed 12 May 2009).

La Motta, J. (1970) *Raging Bull*, with Joseph Carter and Peter Savage. Englewood Cliffs, NJ: de Capo Press.

Laqueur, T. (1990) *Making Sex: Body and Gender from the Greeks to Freud*. Cambridge, MA: Harvard University Press.

Laws, S. (1990) *Issues of Blood*. London: Macmillan.

Leifer, E.M. (1995) *Making the Majors: The Transformation of Team Sports in America*. London: Harvard University Press.

Leighton, T. (2007) 'Cost-Cutting Charlton Scrap Women's Team', *The Independent*, 25 June, p. 6.

Lloyd, M. (2007) *Judith Butler*. Cambridge: Polity.

Loy, J.W. (1968) 'The Nature of Sport: A Definitional Effort', *Quest*, 10(1), pp. 1–15.

Mailer, N. ([1975] 1991) *The Fight*. Harmondsworth: Penguin.

Major, J. (2007) *More Than a Game*. New York: Harper Collins.

Mandela, N. (2007) http://www.daylife.com/quote/0fddcbedx9epl?q=Nelson+Mandela (last accessed 6 April 2009).

Mandelbaum, M. (2004) *The Meaning of Sports: Why Americans Watch Baseball; Football and Basketball and What they see When They Do*, Cambridge: Public Affairs, Perseus Book Group.

Mangan, J.A. (1981) *Athleticism in the Victorian and Edwardian Public School*. Cambridge: Cambridge University Press.

Mangan, J.A. (1986) *The Games Ethic and Imperialism: Aspects of the Diffusion of an Ideal*. London: Frank Cass.

Mangan, J.A. (1992) *The Cultural Bond. Sport Empire, Society*. London: Frank Cass.

Mangan, J.A. (1998a) *The Games Ethic and Imperialism*. London: Viking.

Mangan, J.A. (1998b) 'Sport in Society: The Nordic World and Other Worlds', in Meinander, H. and Mangan, J.A. (eds) *The Nordic World*. London: Frank Cass.

Markula, P. (1995) 'Firm But Shapely, Fit But Sexy, Strong But Thin', *Sociology of Sport Journal*, 12(4), pp. 424–453.

Markula, P. (2003) 'Technologies of the Self', *Sociology of Sport Journal*, 20, pp. 87–107.

Markula, P. (2004) 'Tuning Into One's Self: Foucault's Technologies of the Self and Mindful Fitness', *Sociology of Sport Journal*, 21(3), pp. 302–321.

Markula, P. and Pringle, R. (2006) *Foucault, Sport and Exercise: Power, Knowledge and Transforming the Self*. New York: Routledge.

Marqusee, M. (2005 [2000]) *Redemption Song*. London: Verso.

Martin, E. (1989) *The Woman in the Body*. Buckingham: Open University Press.

Martin, B. (1998) 'Feminsim, Criticism and Foucault', in Diamond, I. and
  Quinby, L. (eds) *Feminism and Foucault: Reflections on Resistance*. Boston, MA:
  Northeastern University Press, pp. 3–20.
Mason, T. (1988) *Sport in Britain: A Social History*, London: Faber and
  Faber.
Mason, T. (1993) 'All the Winners and the Half Times...', *Sports Historian*,
  13, pp. 3–13.
Massumi, B. (1998) *Deleuze, Guattari and the Philosophy of Expression*.
  Minneapolis: University of Minnesota Press.
Massumi, B. (2002) *Parables for the Virtual Movement, Affect, Sensation*. Durham,
  NC: Duke University Press.
Mauss, M. (1973 [1934]) 'Techniques of the Body', *Economy and Society*,
  2, pp. 70–88.
Mauss, M. (1979) 'Body Techniques', in Brewster, B. (ed.) *Sociology and
  Psychology: Essays*. London: Routledge, pp. 95–123.
Mauss, M. (1989 [1938]) 'A Category of the Human Mind: The Notion of Per-
  son, the Notion of Self', in Carrithers, S., Collins, S. and Lukes, S. (eds) *The
  Category of the Person*, trans. W.D. Halls. Cambridge: Cambridge University
  Press pp. 1–25.
McChessney, R.W. (1989) 'Media Made Sport: A History of Sports Coverage
  on the United States', in Wenner, L. (ed.) *Media Sports and Society*. Newbury
  Park, CA: Sage, pp. 49–69.
McIlvanney, H. (1996) *McIlvanney on Boxing*. Edinburgh: Mainstream
  Publishing.
McIntosh, P.C. (1971) *Sport in Society*. London: C.A. Watts.
McKay, J., Messner, M. and Sabo, D. (2000) *Masculinities, Gender Relations and
  Sport*. London: Sage.
McNay, L. (1992) *Foucault and Feminism*. Cambridge: Polity.
McNulty, P. (2007) 'Players Need Protection', http://news.bbcsport1/hi/
  football/7151857stm (last accessed 4 January 2008).
McPherson, B.D., Curtis, J.E. and Loy, J.W. (1989) *The Social Significance of
  Sport*. Champaign, IL: Human Kinetics.
McRae, D. (1996) *Dark Trade: Lost in Boxing*. London: Mainstream Publishing.
Mennesson, C. (2000) '"Hard" Women and "soft" Women', *International
  Review for the Sociology of Sport*, 35(1), pp. 21–33.
Merleau-Ponty, M. (1962) *Phenomenology of Perception*. New York: Routledge.
Messner, M. (2002) *Taking the Field: Women, Men and Sports*. Minneapolis:
  University of Minnesota Press.
Messner, M. and Sabo, D.F. (eds) (1994) *Sex Violence and Power in Sports*.
  Freedom, California: Crossing Press.
Messner, M. and Sabo, D. (2000) *Masculinities, Gender Relations and Sport*.
  London: Sage.
Miller, T., Lawrence, G., McKay, J. and Rowe, D. (2001) *Globalization and Sport:
  Playing in the World*. London: Sage.
Mitchell, K. (2003) *War Baby: The Glamour of Violence*. London: Yellow Jersey
  Press.

Mohanty, C.T., Russo, A. and Torres, L. (eds) (1991) *Third World Women and the Politics of Feminism*. Bloomington and Indianapolis: Indiana University Press.

Moi, T. (1987) *French Feminist Thought*. Oxford: Blackwell.

Moi, T. (1999) *What is a Woman and Other Essays*. Oxford: Oxford University Press.

Moi, T. (2000) *What is a Woman?* Oxford: Oxford University Press.

Morgan, D. (1992) *Discovering Men*. London: Routledge.

Modood, T. (2007) *Multiculturalism*. Cambridge: Polity.

Mossop, J. (1997) 'Lewis Looks Forward to a Bout of Unification After McCall Fiasco', *Sunday Telegraph*, 9 February, S7.

Mulvey, L. (1975) 'Visual Pleasure and Narrative Cinema', *Screen*, Autumn, 16(3), pp. 6–18.

Nandy, A. (1989) *The Tao of Cricket: On Games of Destiny and the Destiny of Games*. New Delhi: Oxford University Press.

Norridge, J. (2008) *Can We Have Our Ball Back, Please? How the British Invented Sport*. London: Allen Lane.

Noyes, J.K. (2002) 'Nature, History and the Failure of Language', in Goldberg, D.T. and Quayson, A. (eds) *Relocating Postcolonialism*. Oxford: Blackwell, pp. 270–281.

Oakley, A. (1972) *Sex, Gender and Society*. London: Maurice Temple Smith.

Oakley, A. (1976) *Sex, Gender and Society*. London: Gower.

Oates, J.C. (1987) *On Boxing*. London: Bloomsbury.

O'Connor, D. (ed.) (2002) *Iron Mike: A Mike Tyson Reader*. New York: Thunder's Mouth Press.

O' Neill, J. (1985) *Five Bodies: The Human Shape of Modern Society*. Ithaca, NY: Cornell University Press.

Orbach, S. (1985) *Fat is a Feminist Issue*. London: Faber.

Osborne, P. and Standford, S. (eds) (2002) *Philosophies of Race and Ethnicity*. London: Continuum.

Pathak, P. (2008) 'Making a Case for Multiculture: From the "Politics of Piety" to the Politics of the Secular', *Theory Culture and Society*, 25(5), pp. 123–141.

Parekh, B. (2000) *The Future of Multi ethnic Britain*. The Parekh Report, London: Profile Books, The Runneymede Trust.

Panorama (2006) http://news.bbc.co.uk/1/hi/programmes/panorama/5219906.stm (last accessed 12 November 2006).

Penley, C., Lyon, E., Spigel, L. and Bergstrom, J. (eds) (1991) *Close Encounters: Film, Feminism and Science Fiction*. New York and London: Verso.

Perkin, H. (1989) 'Teaching the Nations How to Play: Sport and Society in the British Empire and Commonwealth', *Journal of International Sports*, 6(2), pp. 145–155.

Petersen, R. (1992) *Only the Ball was White*. New York: Oxford University Press.

PAT 10 (1999) *National Strategy for Neighbourhood Renewal: Policy Action Team Audit: Report of the Policy Action Team 10: The Contribution of Sport and the Arts*. London: DCMS.

Pateman, C. (1989) 'Feminist Critiques of the Public/Private Dichotomy', in Phillips, A. (ed.) *Feminism and Equality*. Oxford: Basil Blackwell, pp. 103–126.

Phelps, M. (2008) http://www.michaelphelps.com/2004/english.html (accessed 25 November 2008).

Popplewell, Lord Justice (1986) *Inquiry into Crowd Safety at Sports Grounds*, Command Paper 9710. London: HMSO.

Poovey, M. (1988) 'Feminism and Deconstructivism', *Feminist Studies*, 14, pp. 51–65.

Price, M. and Shildrick, J. (1999) Feminist Theory and the Body, Edinburgh, Edinburgh University Press.

Rader, B.G. (1984) *In Its Own Image: How Television Has Transformed Sports*. London: Free Press.

Rail, G. and Harvey, J. (1995) 'Body at Work: Michel Foucault and the Sociology of Sport', *Sociology of Sport Journal*, 12(2), pp. 164–79.

Ramazanoglu, C. and Holland, J. (2005) *Feminist Methodology: Challenges and Choices*. London: Sage.

Rawls, J. (1972) *A Theory of Justice*. Oxford: Oxford University Press.

Rees, T. (2002) 'The Politics of "Mainstreaming" ', in Breitenbach, E., Brown, A., MacKay, F. and Webb, J. (eds) *The Changing Politics of Gender Equality in Britain*. Basingstoke: Palgrave Macmillan.

Rein, I., Kotler, P. and Stoller, M. (1997) *High Visibility: The Making and Marketing of Professionals into Celebrities*. Chicago: NTC Business Books.

Renson, R. (1998) 'The Cultural Dilemma of Traditional Games', in Duncan, M.C., Chick, G. and Aycock, A. (eds) *Play and Culture Studies*. Westport: Greenwood Publishing Group, pp. 51–58.

Remnick, D. (1998) *King of the World: Muhammad Ali and the Rise of the American Hero*. New York: Random House.

Richardson, D. (2008) 'Conceptualizing Gender', in Richardson, D. and Robinson, V. (eds) *Introducing Gender and Women's Studies*, 3rd edition. London: Palgrave Macmillan, pp. 3–19.

Riddell, D. (2008) http://edition.cnn.com/2008/SPORT/05/16/athletics.pistorius/index.html?section=cnnl (last accessed 4 December 2008).

Riesman, D. and Denney, R. (1951) 'Football in America: A Study in Cultural Diffusion', *American Quarterly*, 3, pp. 309–325.

Rinehart, R. (2005) ' "Babes" and Boards. Opportunities in New Millennium Sport?' *Journal of Sport & Social Issues*, 29(3), pp. 232–255.

Rinehart, R. and Sydnor, S. (eds) (2003) *To the Extreme: Alternative Sports, Inside and Out*. Albany: New York Press.

Rojek, C. (2001) *Celebrity*. London: Reaktion Books.

Rose, N. (1996a) *Inventing Our Selves*. Cambridge: Cambridge University Press.

Rose, N. (1996b) 'The Death of the Social? Re-Figuring the Territory of Government', *Economy and Society*, 25(3), August, pp. 327–356.

Rose, N. (1999 [1989]) *Governing the Soul: The Shaping of the Private Self*. London: Free Association Books.

Rojek, C. (2000) *Leisure and Culture*. London: Macmillan.

Roche, M. (2006) 'Mega Events and Modernity Re-visited: Globalization and the Case of the Olympics', *Sociological Review*, 54(2), pp. 25–40.

Rowe, D. (1999) *Sport, Culture and the Media: The Unruly Trinity*. Buckingham: Open University Press.

Sage, G. (1998) *Power and Ideology in American Sport*, 2nd edition. Champaign, IL: Human Kinetics.

Said, E. (1978) *Orientalism*. Harmondsworth: Penguin.

Sammons, J. (1988) *Beyond the Ring: The Role of Boxing in American Society*. Chicago: University of Chicago Press.

Saner, E. (2008) 'The Gender Trap', http://www.guardian.co.uk/sport/2008/jul/30/olympicgames2008.gender (last accessed 7 December 2008).

Scannell, P. and Cardiff, D. (1991) *A Social History of British Broadcasting, 1922–1938*. Oxford: Basil Blackwell.

Scarry, E. (1985) *The Body in Pain: The Making and Unmaking of the World*. New York: Oxford University Press.

Scorsese, M. (1980) *Raging Bull*.

Scott, R. (1982) *Blade Runner*.

Scott, J. (1997) 'Deconstructing Equality-versus Difference: Or, the Uses of Post-structuralist Theory for Feminism', in Tietjens, M.D. (ed.) *Feminist Social Thought: A Reader*. London: Routledge, pp. 757–770.

Schwarz, B. (1996) 'Unspeakable Histories', in Osborne, P. and Standford, S. (eds), pp. 81–96.

Segel, H.G. (1999) *Body Ascendant*. Baltimore: John Hopkins University Press.

Segal, L. (1990) *Straight Sex: The Politics of Pleasure*. London: Virago.

Segal, L. (1994) *Slow Motion: Changing Masculinities Changing Men*. London: Virago.

Shaw, J. (2005) 'Mainstreaming Equality and Diversity in the European Union', *Current Legal Problems*, 58, pp. 255–312.

Shilling, C. (1993) *The Body and Social Theory*. Sage: London.

Shusterman, R. (1992) *Pragmatist Aesthetics: Living beauty, Rethinking Art*. Oxford: Blackwell.

Shusterman, R. (2000) *Performing Live: Aesthetic Alternatives for the Ends of Art*. Ithaca, NY: Cornell University Press.

Skeggs, B. (1997) *Formations of Class and Gender*. London: Sage.

Singh, V.J. (2009) http://en.wikipedia.org/wiki/Vijay_Singh (last accessed 26 January 2009).

Smart, B. (2005) *The Sport Star*. London: Sage.

Smith, D.E. (1997) 'Comments on Hekman's "Truth and Method: Feminist Standpoint Theory Re-Visited', *Signs*, 22(21), pp. 392–397.

Smith, D.E. (1990) *Texts Facts and Femininity*. London: Routledge.

Smith, E. (2008) *What Sport Tells us About Life; Bradman's Average, Zidane's Kiss and Other Sporting Lessons*. London: Viking.

Smith, R.R.R. (1991) *Hellenistic Sculpture*. London: Thames and Hudson.

Sobchack, V. (2004) *Carnal Thoughts: Embodiment and Moving Image Culture*. Berkeley: University of California Press.

Sobchack, V. (1999 [1987]) 'Postfuturism', in Kirkup, G., Janes, L., Woodward, K. and Hovendon, F. (eds) *The Gendered Cyborg*. London: Routledge, pp. 136–147

Sociology of Sport Journal (2001) Special issue on disability and sport, 18(1). Online. www.humankinetics.com or www.sagepub.co.uk.

Spence, J. (2008) 'The Body Beautiful', Lecture 4, *The Reith Lectures*, London: BBC, http://www.bbc.co.uk/radio4/reith2008/ (last accessed 6 April 2009).

Squires, J. (1999) *Gender and Political Theory*. Cambridge: Polity.

Stallybrass, P.D. and White, A. (1986) *The Politics and Poetics of Transgression*. London: Methuen.

Stanley, L. (1984) 'Should "Sex" Really be "Gender" or "Gender" Really be "Sex"?', in Anderson, R. and Sharrock, W. (eds) *Applied Sociology*. London: Allen and Unwin.

Stanley, L. (ed.) (1997) *Knowing Feminisms: On Academic Borders, Territories and Tribes*. London: Sage.

Stanley, L. and Wise, S. (1993) *Breaking Out Again: Feminist Ontology and Epistemology*. London: Routledge.

Sugden, J. (1996) *Boxing and Society: An International Analysis*. Manchester: Manchester University Press.

Szenassy, S. (1995) *Independent on Sunday*, 16 July.

Taylor, Lord Justice (1990) *The Hillsborough Stadium Disaster: Inquiry Final Report*. London: HMSO.

Taylor, M. (1992) 'Formal versus Informal Incentive Structures and Legislative Behavior: Evidence from Costa Rica', *Journal of Politics*, 54(4), pp. 1055–1073.

Taylor, N. (2004) 'Can Football Clubs and Their Communities Co-Exist?', in Wagg, S. (ed.) *British Football and Social Exclusion*. London: Routledge, pp. 47–66.

Tomlinson, A. (1996) 'Olympic Spectacle: Opening Ceremonies and Some Paradoxes of Globalization', *Media, Culture and Society*, 18(4), pp. 583–602.

Tomlinson, A. (2002) 'Theorizing Spectacle: Beyond Debord', in Sugden, J. and Tomlinson, A. (eds) *Power Games: A Critical Sociology of Sport*. London: Routledge, pp. 44–60.

Tomlinson, A. (2007) *The Sports Studies Reader*. London: Routledge.

Thompson, C.S. (2006) *The Tour de F 2rance: A Cultural History*. Berkeley, CA: University of California Press.

Tuner, B. (1996) *The Body and Society*, 2nd edition. London: Sage.

Turner, B.S. (1984) *The Body and Society: Explorations in Social Theory*. Oxford: Blackwell.

Turner, B.S. (1994) 'Introduction to C. Buci-Glucksmann Baroque Reason', *The Aesthetics of Modernity*. London: Sage, pp. 1–36.

Turner, G. (2004) *Understanding Celebrity*. London: Sage.

Van der Lippe, G. (1994) *International Review for the Sociology of Sport*, 29(2), pp. 211–231

Wacquant, L. (1993) 'Positivism', in Outhwaite, W. and Bottomore, T. (eds) *The Blackwell Dictionary of Twentieth Century Thought*. Oxford: Blackwell.

Wacquant, L. (2001) 'Whores, Slaves and Stallions: Languages of Exploitation and Accommodation among Professional Fighters', *Body and Society*, 7, pp. 181–194.

Wacquant, L. (2004) *Body and Soul: Notebooks of an Apprentice Boxer*. Oxford: Oxford University Press.

Wacquant, L. (1995a) 'Pugs at Work: Bodily Capital and Bodily Labour among Professional Boxers', *Body and Society*, 1(1), pp. 65–93.

Wacquant, L. (1995b) Review article, 'Why Men Desire Muscles', *Body and Society*, 1, pp. 163–179.

Wacquant, L. (1995c) 'The Pugilistic Point of View: How Boxers Think About Their Trade', *Theory and Society*, 24(4), pp. 489–535.

Wade, P. (2002) *Race, Nature and Culture*. London: Pluto Press.

Wagg, S. (ed.) (2004) *British Football and Social Exclusion*. London: Routledge.

WBA (2008) http://www.wbaonline.com/ (last accessed 12 October 2008).

Women's Football (2009) http://www.thefa.com/Womens/ (last accessed 6 April 2009).

World Boxing Council (2008) http://www.wbcboxing.com/WBCboxing/portal/docs/word/NEW_RULES_AND_REGULATIONS_WEB%20final.doc (last accessed 12 October 2008).

Walsh, A. and Giulianotti, R. (2006) *Ethics, Money and Sport*. London: Routledge.

Walsh, A.J. and Giulianotti, R. (2001) This Sporting Mammon: A Normative Critique of the Commodification of Sport, *Journal of the Philosophy of Sport*, 28(1), pp. 53–77.

Werbner, P. (2002) *Imagined Diasporas among Manchester Muslims*. Santa Fe New Mexico: School of American Research Press.

Walkerdine, V. (1986) 'Video Replay: Families, Films and Fantasies', in Burgin, V., Donald, J. and Kaplan, C. (eds) *Formations of Fantasy*. London: Methuen.

WBAN (2005) 'Women's Boxing Archive Network', http://www.women-boxing.com (last accessed 10 November 2006).

Weber, M. (1978) *Economy and Society*. Berkeley, CA: University of Berkeley Press.

Whannel, G. (2002) *Media Sport Stars: Masculinities and Moralities*. London: Routledge.

Wheaton, B. (2002) 'Babes on the Beach, Women in the Surf', in Tomlinson, A. and Sugden, J. (eds) *Power Games: A Critical Sociology of Sport*. London: Routledge.

Wheaton, B. (2004) *Understanding Lifestyle Sport: Consumption, Identity and Difference*. London: Routledge.

Wheaton, B. and Beal, B. (2003) *Keeping It Real: Subcultural Media and the Discourses of Authenticity in Alternative Sport*, 38(2), pp. 155–176.

Whyte, W. (1955) *Street Corner Society*. Chicago: University of Chicago Press.

Williams, J. and Wagg, S. (1991) *British Football and Social Change*. Leicester: Leicester University Press.

Williams, J. and Woodhouse, J. (1991) 'Can Play, Will Play? Women and Football in Britain', in Williams, J. and Wagg, S. (eds) *British Football and Social Change: Getting into Europe.* Leicester: Leicester University Press.

Williamson, M. (2009) http://content-www.cricinfo.com/ci/content/story/248600.html (last accessed 24 January 2009).

Women's FA (2008) www.thefa.com/womens (last accessed 5 January 2008).

Woodward, K. (1997) *Whose Body?* BBC television programme: Open University/BBC.

Woodward, K. (2004) 'Rumbles in the Jungle. Boxing; Racialization and the Performance of Masculinity', *Leisure Studies,* 23(1), pp. 1–13.

Woodward, K. (2007a) *Boxing, Masculinity and Identity: The "I" of the Tiger.* London: Routledge.

Woodward, K. (2007b) 'On and Off the Pitch: Diversity Policies and Transforming Identities', *Cultural Studies.* 21(4/5), pp. 758–778.

Woodward, K. (2008a) 'Gendered Bodies, Gendered Lives', in Richardson, D. and Robinson, V. (eds) *Introducing Gender and Women's Studies,* 3rd edition. Basingstoke: Palgrave Macmillan, pp. 75–90.

Woodward, K. (2008b) 'Hanging Out and Hanging About: Insider/Outsider Research in the Sport of Boxing', *Ethnography,* 9(4), pp. 536–560.

Woodward, K. (2009) *Bodies on the Margins: Regulating bodies, regulatory bodies Leisure Studies,* 28(2), pp. 143–156.

Wrench, J. (2005) 'Diversity Management Can be Bad For You', *Race and Class,* 46(3), p. 7384.

Young, I.M. (1990) *Throwing Like a Girl and Other Essays on Feminist Philosophy and Social Theory.* Bloomington and Indianapolis: Indiana University Press.

Young, I.M. (2005a) 'Throwing Like a Girl', in Young, I.M. (ed.) *On Female Body Experience.* Oxford: Oxford University Press, pp. 3–26.

Young, I.M. (2005b) *On Female Body Experience. "Throwing Like a Girl and Other essays".* Oxford: Oxford University Press.

Žižek, S. (1992) *Enjoy Your Symptom! Jacques Lacan in Hollywood and Out.* London: Routledge.

Žižek, S. (1993) *Tarrying with the Negative: Kant, Hegel and the Critique of Ideology.* Durham, NC: Duke University Press.

Žižek, S. (1997) 'Multiculturalism, or, the Cultural Logic of Multinational Capitalism', *New Left Review,* 225, September/October, pp. 28–51.

Žižek, S. and Dolar, M. (2002) *Opera's Second Death.* London: Routledge.

Žižek, S., Butler, J. and Laclau, E. (2000) *Contingency, Hegemony, Universality: Contemporary Dialogues on the Left.* London and New York: Verso.

Index